T0393204

SYSTEM INNOVATION IN A POST-PANDEMIC WORLD

Smart Science, Design and Technology

ISSN: 2640-5504
eISSN: 2640-5512

Book Series Editors

Stephen D. Prior
Faculty of Engineering and Physical Sciences
University of Southampton
Southampton, UK

Siu-Tsen Shen
Department of Multimedia Design
National Formosa University
Taiwan, R.O.C.

VOLUME 5

PROCEEDINGS OF THE IEEE 7TH INTERNATIONAL CONFERENCE ON APPLIED SYSTEM INNOVATION (ICASI 2021), SEPTEMBER 24–25, 2021, ALISHAN, TAIWAN

System Innovation in a Post-Pandemic World

Editors

Artde Donald Kin-Tak Lam
Fujian University of Technology, P.R. China

Stephen D. Prior
Aeronautics, Astronautics and Computational Engineering, Faculty of Engineering and Physical Sciences, The University of Southampton, Southampton, United Kingdom

Sheng-Joue Young
National United University, Miaoli District, Taiwan

Siu-Tsen Shen
Department of Multimedia Design, National Formosa University, Taiwan

Liang-Wen Ji
Department of Electra-Optical Engineering, National Formosa University, Huwei District, Taiwan

CRC Press
Taylor & Francis Group
Boca Raton London New York Leiden

CRC Press is an imprint of the
Taylor & Francis Group, an **informa** business

A BALKEMA BOOK

CRC Press/Balkema is an imprint of the Taylor & Francis Group, an informa business

© 2022 selection and editorial matter, Artde Donald Kin-Tak Lam, Stephen D. Prior, Sheng-Joue Young, Siu-Tsen Shen & Liang-Wen Ji; individual chapters, the contributors

Typeset by MPS Limited, Chennai, India

The right of Artde Donald Kin-Tak Lam, Stephen D. Prior, Sheng-Joue Young, Siu-Tsen Shen & Liang-Wen Ji to be identified as the authors of the editorial material, and of the authors for their individual chapters, has been asserted in accordance with sections 77 and 78 of the Copyright, Designs and Patents Act 1988.

Library of Congress Cataloging-in-Publication Data
A catalog record has been requested for this book

First published 2022
Published by: CRC Press/Balkema
                Schipholweg 107C, 2316 XC Leiden, The Netherlands
                e-mail: enquiries@taylorandfrancis.com
                www.routledge.com – www.taylorandfrancis.com

ISBN: 978-1-032-24392-4 (HBk)
ISBN: 978-1-032-24411-2 (Pbk)
ISBN: 978-1-003-27847-4 (eBook)
DOI: 10.1201/9781003278474

Table of contents

Preface

We have great pleasure in presenting this conference proceeding for technology applications in engineering science and mechanics from the selected articles of the International Conference on Applied System Innovation (ICASI 2021), organized by the International (Taiwanese) Institute of Knowledge Innovation and the IEEE, 24–25 September, 2021 at Alishan, Taiwan.

The ICASI 2021 conference was a forum that brought together users, manufacturers, designers, and researchers involved in the structures or structural components manufactured using smart science. The forum provided an opportunity for exchange of the research and insights from scientists and scholars thereby promoting research, development and use of computational science and materials. The conference theme for ICASI 2021 was "System Innovation in a Post-Pandemic World" and tried to explore the important role of innovation in the development of the technology applications, including articles dealing with design, research, and development studies, experimental investigations, theoretical analysis and fabrication techniques relevant to the application of technology in various assemblies, ranging from individual to components to complete structure were presented at the conference. The major themes on technology included Material Science & Engineering, Communication Science & Engineering, Computer Science & Engineering, Electrical & Electronic Engineering, Mechanical & Automation Engineering, Architecture Engineering, IOT Technology, and Innovation Design. About 200 participants, representing 11 countries came together for the 2021 conference and made it a highly successful event. We would like to thank all those who directly or indirectly contributed to the organization of the conference.

Selected articles presented at the ICASI 2021 conference will be published as a series of special issues in various journals. In this conference proceeding we have some selected articles on various themes. A committee consisting of experts from leading academic institutions, laboratories, and industrial research centres was formed to shortlist and review the articles. The articles in this conference proceeding have been peer reviewed to the usual standards. We are extremely happy to publish this conference proceeding and dedicate it to all those who have made their best efforts to contribute to this publication.

Professor Siu-Tsen Shen & Dr Stephen D. Prior

System Innovation in a Post-Pandemic World – Kin-Tak Lam et al. (Eds)
© 2022 Copyright the Editor(s), ISBN: 978-1-032-24392-4

Editorial Board

System Innovation in a Post-Pandemic World – Kin-Tak Lam et al. (Eds)

A study on the influence of building agricultural biomass power plants on air quality in Taiwan: Taking Yunlin area as an example

Chuan-Hsi Shih*
Ph.D. Program of Mechanical and Energy Engineering, Kun Shan University, Tainan, Taiwan

Huann-Ming Chou
Department of Mechanical Engineering, Kun Shan University, Tainan, Taiwan

ABSTRACT: This study investigated the impact of air pollution on the health of people in Taiwan. In addition, the coefficient method was adopted to compare the amount of pollutants released from the burning of various agricultural wastes "into the open air" with that produced from burning agricultural wastes "directly in the biomass power plant," exploring the impact of the agricultural waste treatment method on air pollution in the Yunlin area of Taiwan. The findings of this study can serve as a useful reference for the government in the development of biomass energy.

Keywords: Agricultural waste, Air pollution, Biomass power plant

1 INTRODUCTION

Taiwan relies on imports for almost all of its energy supply to support livelihoods and industry, although rice yields are very high, enough not only to meet the local demand but also to export and earn foreign exchange. The traditional methods of rice straw disposal generally encompass open burning, composting, burying, and other methods. Of these, the microbial burial method for straw decomposition is a policy that has been widely promoted by the Taiwan government in recent years. However, on-site burial takes at least 20 days for complete decomposition; incomplete on-field straw decomposition due to any errors in the process may cause suffocation disease in the second crop rice seedlings. Therefore, most farmers tend to adopt the burning method during this period (H.J. Tang, 2018). In terms of the straw decomposition rate, on-site burning is the most time-efficient approach. However, this approach causes air pollution and negatively affects the respiratory health of residents of surrounding areas. According to a 2019 report by the International Energy Agency, global greenhouse gas emissions arising from energy use surged to a record high of 33.1 billion tons in 2018, and the growth rate of global energy consumption was also twice the average growth rate since 2010 (Global Energy & CO_2 Status Report 2019, IEA, Paris). Humanity is facing a crisis of environmental destruction and gradual depletion of energy resources. In this context, the development of

clean alternative energy has become an important mission for the third industrial revolution, and is also an important task for Taiwan's current energy development programs. Biomass can be converted into energy sources, and if it can be efficiently used as an alternative energy source, the dependence on fossil energy can be decreased, which will have positive effects on controlling the greenhouse effect. In circumstances where the supply of agricultural biomass residues can be ensured, the Taiwanese government should actively consider the development of bioenergy from waste straw and corn stalks. This not only would solve the problem of agricultural waste removal and disposal but would also reduce air pollution caused by open burning and offer a solution to this age's energy shortage.

2 LITERATURE REVIEW

Open burning involves incomplete combustion and an inability to manage harmful pollutants; it often releases SO_2, CO, NOx, PM10, and NMHCs into the air. Chih-Yuan Lin (2004) analyzed the air quality monitoring data in the Yunlin and Chiayi regions from November 27 to 28, 1991, and found that on November 27, five counties in central and southern Taiwan had total emissions of CO, NMHCs, NOx, PM10, and SO_2 of 665.4, 153.2, 40.5, 103.5, and 10.7 metric tons, respectively, implying that serious air pollution in these areas was a result of burning farm waste (C.Y. Lin, 2004). As indicated by the research of Jing-Liang Wang (2000), organic matter accounts for about

---

*Corresponding Author

90% of the composition of rice straw. If the biomass is not completely burned during the combustion process, polycyclic aromatic hydrocarbons (PAHs) are produced. Because of the special topography and climate of Taiwan, pollutants produced by open burning are generally not easily diffused, and thus suspended particulates and high concentrations of PAHs are found in some areas (J.L. Wang, 2000). When rice straw is being intensively burned, the average concentration of TSP in the atmosphere reaches 254 $\mu g/m^3$, much higher than the value of 108 $\mu g/m^3$ during the nonintensive burning season (Y.L. Su, 2006). The United States has classified such substances as carcinogens. Some PAHs have been proven to cause cancer if they accumulate in the human body for a long period of time, and are harmful to health (Carl E. Cerniglia, 1992). The substances NOx and NMHCs produced during open burning form the secondary pollutant O_3 under conditions that facilitate photochemical reactions. With poor diffusion and unfavorable weather conditions coupled with transmission effects, even cross-regional air pollution occurs (C.H. Chen, 2004). Straw contains 18.67% ash, 74.67% silicon dioxide, 15.86% fixed carbon, 3.01% calcium oxide (CaO), 1.75% magnesium oxide (MgO), 0.96% nitrogen dioxide (Na_2O), and 12.30% potassium dioxide (K_2O). Straw containing these harmful elements will eventually release 70% carbon dioxide (CO_2), 7% carbon monoxide (CO), 0.66% methane (CH_4), and 2.09% nitrogen dioxide (N_2O), as well as PM10 and PM2.5 particulate matter, into the environment when burned, which poses a threat to human health. A study shows that PM10 and PM2.5 re-leases increase the incidence of acute respiratory infections in children living in surrounding areas (Charu Batra, 2017). Another study points out that particulate matter pollution caused by burning straw increases cardiopulmonary mortality in humans, and estimates that mortality rates will increase by 3.25% if PM2.5 is increased by 10 $\mu g/m^3$ (G. He et al., 2020).

3 METHODOLOGY

According to 2018 agricultural statistics of the Executive Yuan's Council of Agriculture, crops grown in Yunlin County include rice, corn, sorghum, soybeans, peanuts, and sugarcane, with a total planting area of 69,952 hectares and a total production of 589,673 metric tons. Previous studies calculated the coefficients for biomass/crop ratios of rice, corn, sorghum, soybeans, peanuts, and sugarcane, which were, respectively, 2.024 (rice straw 1.757, rice husks 0.267), 2.473 (corn stalks 2.0, corn cob 0.273), 1.25 (sorghum stalks 1.25), 3.5 (soybean stalks and pods), 2.777 (peanut stalks 2.3, peanut shells 0.477), and 0.39 (sugarcane bagasse 0.29, sugarcane leaves 0.1) (A. Koopmans et al., 1997). Based on these figures, the total residual quantity of crops discussed in this study was estimated to be approximately 921,798.65 metric tons. Another study indicates that about 17.2% of crop residues are directly backfilled in farmland, about 21.5% are used as feed, about 14% are harvesting losses, about 2.9% are used as industrial raw materials, about 24% are used for bioenergy, and about 20.5% are released into the air through open burning (Y.J. Wang et al. 2010). This study assumed biomass being transported to the power plant and then directly burned (Scenario 1). In this process, 14% was harvesting loss, 17.2% was directly backfilled in farmland, and the remaining 68.8% was used as a source of biomass for power generation. Thus, it was estimated that the 2018 output of agricultural waste in Yunlin County was about 634,198 metric tons. It was also assumed that if the approach of on-site open burning was adopted (Scenario 2), in which the total collected amount would not list the "harvesting loss" and "backfilled in farmland" categories, then the estimated amount of open burning would be approximately 921,798.65 metric tons (see Table 1).

The air pollutant emission coefficients (kg pollutants/ton agricultural waste) used in this study were

Table 1. Estimated amount of 2018 crop residue biomass in Yunlin County.

Crop	Area (hectares)	Amount (metrictons)	Biomass/ cropratio	Residue biomass	
				Burning in power plant (Scenario 1)	Farmland on-site burning (Scenario 2)
Rice	44,652	286,426	2.024 (Rice straw 1.757 Rice husks 0.267)	398,852	579,726.22
Corn	6,626	54,071	2.473 (Corn stalks 2.0 Cobs 0.273)	91,998	133,717.58
Sorghum	1	4	1.25 (Sorghum stalks)	3	5
Soybeans	164	265	3.5 (Soybean stalks, pods)	638	927.5
Peanuts	15,616	46,229	2.777 (Peanut stalks 2.3 Peanut shells 0.477)	88,324	128,377.93
Sugarcane	2,893	202,678	0.39 (Sugarcane bagasse 0.29 Sugarcane leaves 0.1)	54,383	79,044.42

sourced from the study done by Lu and Zhang (2010), which investigated pollutant emissions related to agricultural biomass waste used for bioenergy in the processes of transportation, drying/crushing, and combustion for power generation, as shown in Table 2 (Lu & Zhang, 2010). A diesel vehicle was considered in the "Transportation" section of the table, with a diesel heating value of 42.6 MJ/kg, a carrying capacity of 16 tons per load, and an average transport distance of 90 km. Since Yunlin, the subject of this study, has a relatively small area, the transport distance was assumed to be 20 km. Multiplying this coefficient by the agricultural biomass waste, the pollutant emissions from agricultural biomass waste when used for bioenergy (Scenario 1) during the processes of production/harvesting, transportation, and drying/crushing were obtained. As for farmland on-site open burning (Scenario 2), the pollutant emissions were obtained by directly multiplying the total harvested amount by the coefficient (see Table 3).

Coefficient reference source: Koopmans, A., Koppejan, J., 1997. Agricultural and forest residues-generation, utilization and availability. Paper presented at the Regional Consultation on Modern Applications of Biomass Energy 6, 10.

a. Coefficient and emission reference source: Lu and Zhang (2010). Life-cycle implications of using crop residues for various energy demands in China. Environmental Science & Technology, 44(10), 4026–4032.

b. Lu and Zhang assumed a transport distance of 90 km, which was adjusted to 20 km for the case of the Yunlin region in this study. The transportation coefficient was adjusted to 0.222 (=20/90) times the original value.

Emission coefficient reference source: Lu and Zhang (2010). Life-cycle implications of using crop residues for various energy de-mands in China. Environmental Science & Technology, 44(10), 4026–4032.

4 RESULTS

Production and collection procedures should also be considered in the crop production process. However, this study focused on evaluating air pollution arising from the reuse of agricultural biomass waste for power generation, and thus did not incorporate air pollutant emissions during the production and collection stages.

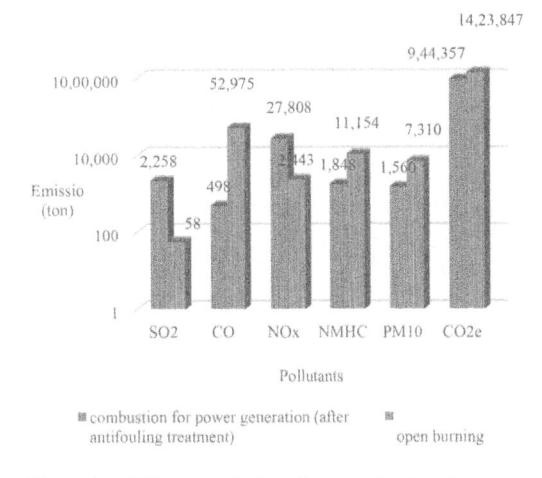

Figure 1. Pollutant emissions from combustion for power generation and open burning
Chart source: summarized for this thesis.

Table 2. On-site burning of agricultural wastes in power plants—combustion pollutant emission coefficients and emissions in the processes of transportation, drying/crushing, and combustion for power generation.

Handling Process	Coefficient (kg/ton) Emission (ton)	SO_2	CO	NOx	NMHC	PM10	CO_2e
Transportation	Coefficient	1.3	0.76	38	2.3	0.62	3.3
	Emission	824	482	24,076	1,436	393	2,067
Drying/	Coefficient	0.882	0	0.615	0	0	118.1
crushing	Emission	559	–	390	–	–	74,898
Combustion for power generation	Coefficient	1.38	0.025	5.27	0.65	1.84	1367.7
(After pollution prevention treatment)	Emission	875	16	3,342	412	11,167	8,67,392
Total Emission		2,258	498	27,808	1848	1,560	9,44,357

Table 3. Estimation of 2018 air pollution emissions from open burning of crop biomass residues in the Yunlin region.

Emission Coefficient (kg/ton)	SO_2	CO	NOx	NMHC	PM_{10}	CO_2e
Open burning coefficient	0.063	57.47	2.65	12.1	7.93	1544.64
Emission (tons)	58	52,975	2,443	11,154	7,310	1,423,847

3

After summarizing the relevant emission coefficients, the results were shown for the calculated gas emissions under Scenario 1 (in which agricultural waste is transported to the power plant for direct combustion after the processes of transportation, drying/crushing, and combustion for power generation) as well as Scenario 2 (in which the approach of farmland on-site open burning is adopted); the released SO_2, CO, NO, NMHCs, PM10, and CO_2 emissions were then calculated. The results were summarized as follows: Except for SO_2 and NOx, the total pollutant emissions of different kinds of agricultural waste calculated for open burning were higher than those of direct burning in a power plant. Emissions of CO, NMHC, PM10 and CO_2 produced by open burning were respectively 106 times, 6 times, 4.69 times and 1.5 times the emissions produced by direct burning in a power plant (see Figure 1) This shows that open burning of agricultural waste poses a real threat to human health. If a bioenergy power plant could be set up in the Yunlin region, and air pollution control equipment installed to treat emissions in accordance with the national standards before being released into the atmosphere, emissions of many pollutants would be reduced.

REFERENCES

Charu Batra, 2017. Stubble Burning in North-West India and its Impact on Health, Research Scholar, Centre for Research in Rural and Industrial Development. Sector 19-A, 160 019, Chandigarh, India.

Carl E. Cerniglia, 1992. Biodegradation of polycyclic aromatic hydrocarbons, Biodegradation 3: 351–368.

C.H. Chen, 2004. Effects of open burning of straw in Taichung, Changhua, Yunlin, Chiayi, and Tainan on the formation of photochemical smog in Kaohsiung and Pingtung, Graduate Institute of Environmental Engineering College of Engineering National Taiwan University Master Thesis, Taipei, 2004.

Global Energy & CO_2 Status Report 2019, IEA, Paris.

G. He, T. Liu, M. Zhou, 2020. Straw Burning, PM2.5, and Death: Evidence from China, Journal of Development Economics, Volume 145, 102468¡C

A. Koopmans, J. Koppejan, 1997. Agricultural and forest residues-generation, utilization and availability. Paper presented at the regional consultation on modern applications of biomass energy 6, 10.

C.Y. Lin, 2004. Effects of Open burning of Agricultural Waste in Yunlin and Chiayi of Taiwan on Air Quality and Pollutant Emissions, Graduate Institute of Environmental Engineering College of Engineering National Taiwan University Master Thesis, Taipei.

Lu, Zhang, 2010. Life-cycle implications of using crop residues for various energy demands in China. Environmental Science & Technology, 44(10), pp. 4026–4032.

Y.L. Su, 2006. Source identification and size distribution of atmospheric polycyclic aromatic hydrocarbons during rice straw burning period, Department of Environmental Engineering and Management, Chaoyang University of Technology Master Thesis, Taichung.

H.J. Tang, 2018. Application Techniques and Precautions for Using Corruption of Rice Straw to Produce Organic Fertilizers.

J.L. Wang, 2000. Master's Thesis: Analysis of the Composition of Fugitive Dust from Pollution Sources in the Central Air Quality Monitoring Area. Department of Environmental Engineering, National Chung Shing University, Taichung.

Y.J. Wang, Y.Y. Bi, C.Y. Gao, 2010. Collectable Amounts and Suitability Evaluation of Straw Resource in China, Scientia Agricultura Sinica, 43(9), pp. 1852–1859.

System Innovation in a Post-Pandemic World – Kin-Tak Lam et al. (Eds)
© 2022 Copyright the Author(s), ISBN: 978-1-032-24392-4

Evaluation of parameter settings for digital audio sampling specifications

Ching-Chien Liang
Department of Popular Music Industry, Southern Taiwan University of Science and Technology, Yungkang, Tainan, Taiwan

Chian-Fan Liou
Department of Visual Communication Design, Southern Taiwan University of Science and Technology, Yungkang, Tainan, Taiwan

Chao-Chih Huang*
Department of Popular Music Industry, Southern Taiwan University of Science and Technology, Yungkang, Tainan, Taiwan

ABSTRACT: For the application of digital audio sampling technology to audio recording, there are several important parameter settings that must be set appropriately in relevant audio software and hardware interfaces during the initiation of preparation for audio recording. Among these settings, there are two important specifications that seem to be basic but are most often confused whether to consumers or professionals. These parameter specifications include bit depth and sample rate. According to the common parameter settings, two kinds of bit depths, 16-bit and 24-bit, and two kinds of sample rate, 44.1 kHz and 48 kHz, are used for sorting and summarizing suggested appropriate parameter settings for engineers' references and applications in audio recording operations. This research took Avid Pro Tools system, the beloved digital audio workstation that has had been highly used in multimedia productions by the global indicator awards of Grammy Award and Oscars winners, as the research platform, to do the dynamic amplitude scale analysis of the audio recording process established by the settings of 16-bit, 24-bit, 44.1 kHz and 48 kHz interactively, following up with focus group interviews and Likert scale questionnaires for further analysis and research discussions. The conclusions of this research suggest how to set the bit depth and sample rate to audio recording engineers during the initial working processes properly. Thus, proper audio amplitude scales can be established in the early stages of digital audio recording and can be well managed for the subsequent audio processing tasks in audio editing, mixing and mastering.

1 INTRODUCTION

To be immersed in the huge leap of digital multimedia technology, for the applications of digital audio sampling technology of audio recording engineering operations, there are several important parameter specification settings that must be set appropriately in audio software and hardware interfaces during the initiation preparation for audio recording tasks. Among these settings, there are two important parameter specifications that seem to be basic but are most often confused whether to most consumers or professional practitioners. The main audio parameter specification settings are bit depth and sample rate.

The common bit depth specifications of digital audio recording engineering technology include 16-bit and 24-bit. In terms of bit depth, the direct impact of this parameter in digital audio recording engineering is the dynamic range of analog audio converted

to digital audio signal strength. The physical unit of audio signal dynamic range is 'decibel' (abbreviated as dB), which is a common unit for measuring sound intensity (Isnawati, Citra, & Hendry, 2019).

The common sample rate specifications of digital audio recording engineering technology are 44.1 and 48 kHz. In terms of sample rate, the direct impact of this parameter on digital audio recording engineering is the numerical upper limit value of the analog audio signal frequency that can be accessed by digital audio sampling system or the software/hardware of the audio recording system (Grondin, Tang, & Glass, 2020). According to Nyquist theorem, the sample rate specification of the software and hardware of the digital audio sampling or recording engineering system must be set at least twice the upper range of the audio frequency value of the target analog audio signal to be recorded (Cook, 2017).

Based on the common parameter settings in digital audio sampling and recording engineering technology, two settings of bit depth—16-bit and 24-bit—and two

*Corresponding Author

DOI 10.1201/9781003278474-2

settings of sample rate—44.1 kHz and 48 kHz—are used for sorting and summarizing suggested settings of important application parameters to contribute suggestions for audio engineers' references and applications in the actual audio recording studio operation processes.

2 EXPERIMENTAL SECTION

This research project made use of Avid Pro Tools audio system, the beloved digital audio workstation which has had been highly used in audio and video productions by the global indicator awards of Grammy Award and Oscars film award winners, as the research platform, doing the dynamic amplitude scale analysis of the audio recording process established by two settings of bit depth and two settings of sample rate interactively, following up with focus group interviews (Hennink, Kaiser, & Weber, 2019) and Likert scale questionnaires (Aji & Larner, 2017) for further analysis and research discussion purposes.

In this research project, 50 university students, who had participated in audio production tasks for more than five projects in recording studios, with age between 18 and 21 years, were invited as the respondents. The research analyses were based on descriptive statistics (Mishra et al., 2019) and linear statistical analysis (Mao & Monahan, 2017) to explore the psychological differences towards respondents under four combination session settings listed in Table 1, which were applied to four individual Avid Pro Tools digital audio workstations (version: Pro Tools | Ultimate 2020.11) to check for the psychological response differences to 50 respondents.

Table 1. Different combination settings used for surveys.

	Bit Depth: 16-bit	Bit Depth: 24-bit
Sample rate: 44.1 kHz	Session A	Session B
Sample rate: 48 kHz	Session C	Session D

A four-point Likert scale questionnaire, as shown in Table 2, was applied for each session setting to ask the respondents to present their feedback opinions about the impressions of the psychological satisfaction levels with audio dynamic range under four different combination session settings.

Table 2. Four-point Likert scale questionnaire.

	Combination Setting	Strongly Disagree	Disagree	Agree	Strongly Agree
1	Session A				
2	Session B				
3	Session C				
4	Session D				

The audio hardware and software configuration settings and the interview steps were as follows:

(1) The control variables used have been listed as follows: One vocal talent and one electric guitar player were invited for singing and music performance; under the same following conditions, all respondents took interviews in the same recording studio, using the same vocal microphone (Blue, Mouse), same electric guitar (Fender, Stratocaster) and the same audio signal preamplifier (SSL, Alpha Channel).

(2) A set of 1-in-4-out audio signal distributor (Radio Design Labs, RU-MLD4 distribution amplifier) was utilized for distributing the same vocal singing and electric guitar playing audio signals simultaneously, to record into four individual audio tracks in four individual same model audio-to-digital digital-to-audio converter interfaces (Avid, HD I/O) with four same individual model computers (Mac Pro, 2.7 GHz 12-Core Intel Xeon E5 processor, 16 GB memory), whilst the same model headphones (AKG, K240) and the same model monitor speakers (GENELEC, 1031A PM) were used for surveys (Figures 1 and 2).

(3) The audio recording session parameter settings were set to (a) record audio file type: Broadcast Wave Format (BWF) and (b) record track type: one single mono audio track, whilst four combination settings of bit depth/sample rate were applied in each Avid Pro Tools digital audio workstation as follows:

- Session A: 16-bit /44.1 kHz
- Session B: 24-bit /44.1 kHz
- Session C: 16-bit /48 kHz
- Session D: 24-bit /48 kHz

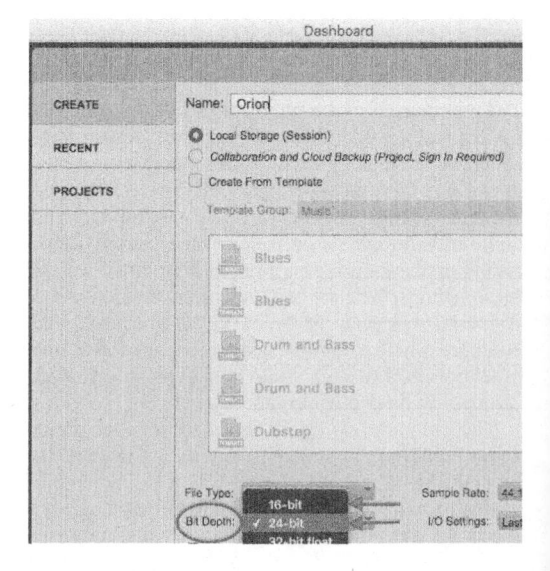

Figure 1. Pro Tools session parameter settings screen, bit depth. (Courtesy of Avid Technology, Inc., www.avid.com.)

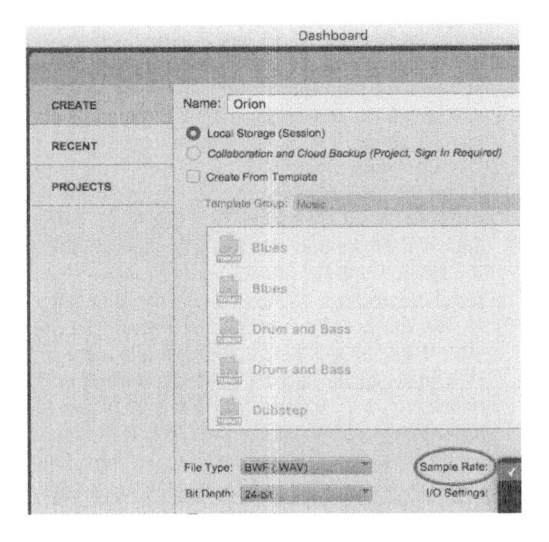

Figure 2. Pro Tools session parameter settings screen, sample rate. (Courtesy of Avid Technology, Inc., www.avid.com.)

(4) The respondents were asked to listen to the same vocal singing and the same electric guitar playing conveyed by the aforementioned headphones and monitor speakers.

(5) Four combination settings of bit depth/sample rate were used for each respondent as the independent variables to check for their opinions about the psychological satisfaction levels with audio dynamic range.

(6) All 50 respondents were asked to finish checking the questionnaires shown in Table 2. The statement question presented in the questionnaire was, 'I *am* satisfied with the audio dynamic range of the playback audio signal.'

3 RESULTS AND DISCUSSION

After the completions of the above investigations and testing processes, the following survey results were obtained:

(1) When combination settings of session bit depth/sample rate were set to 16-bit/44.1 kHz, the answer counts to the question 'I *am* satisfied with the audio dynamic range of the playback audio signal', from 50 respondents were: (a) Strongly Disagree: 25, (b) Disagree: 23, (c) Agree: 2, (d) Strongly Agree: 0.

(2) When combination settings of session bit depth/sample rate were set to 24-bit/44.1 kHz, the answer counts to the question 'I *am* satisfied with the audio dynamic range of the playback audio signal', from 50 respondents were: (a) Strongly Disagree: 0, (b) Disagree: 2, (c) Agree: 20, (d) Strongly Agree: 28.

(3) When combination settings of session bit depth/sample rate were set to 16-bit/48 kHz, the answer counts to the question 'I *am* satisfied with

the audio dynamic range of the playback audio signal', from 50 respondents were: (a) Strongly Disagree: 21, (b) Disagree: 25, (c) Agree: 3, (d) Strongly Agree: 1.

(4) When combination settings of session bit depth/ sample rate were set to 24-bit/48 kHz, the answer counts to the question 'I *am* satisfied with the audio dynamic range of the playback audio signal', from 50 respondents were: (a) Strongly Disagree: 0, (b) Disagree: 1, (c) Agree: 21, (d) Strongly Agree: 28.

According to the two sets of survey data results shown in Figures 3 and 4, respondents showed high level of satisfaction to 24-bit session tasks, no matter the sessions were set in the sample rate of 44.1 or 48 kHz.

In Figures 5 and 6, under different numbers of sample bit, the respondents also showed their higher preferences, as well as were better satisfied with the audio dynamic range to 24-bit session than 16-bit, no matter the sessions were set in the sample rate of 44.1 or 48 kHz.

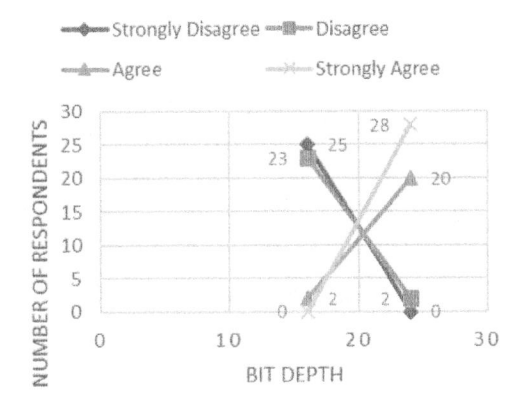

Figure 3. Satisfaction survey at the same sample rate of 44.1 kHz.

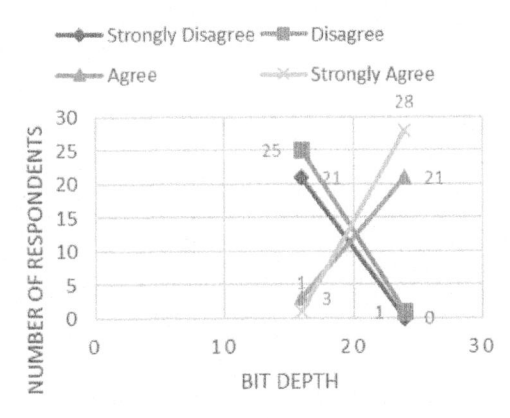

Figure 4. Satisfaction survey at the same sample rate of 48 kHz.

BIT DEPTH: 16-BIT

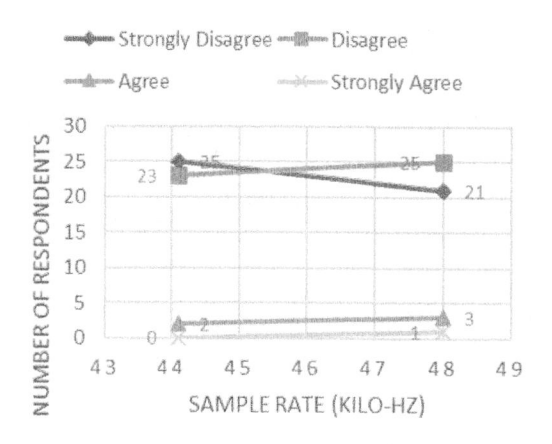

Figure 5.   Satisfaction survey at the same bit depth of 16-bit.

BIT DEPTH: 24-BIT

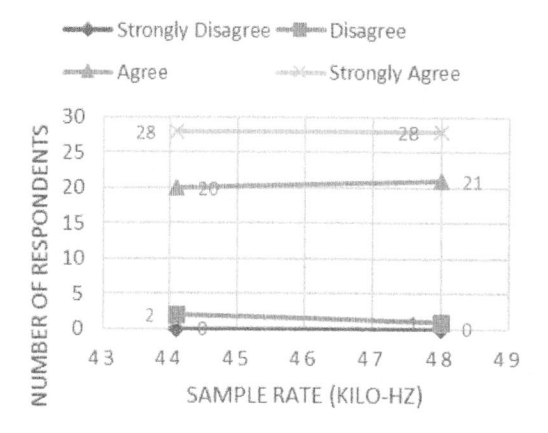

Figure 6.   Satisfaction survey at the same bit depth of 24-bit.

4   CONCLUSION

In this research, four combination settings of bit depth/sample rate were used to check respondents' feedback opinions with the psychological satisfaction levels with audio dynamic range. The following inferences were obtained as the research results.

(1) For the concerned choices of bit depth, the experimental results showed that with the respondents' psychological evaluations to the session setting at 24-bit, the audio dynamic range satisfaction level was much higher than the session setting at 16-bit.

(2) In the audio pre-production stage of the recording tasks, if the recording engineer is not that confident

with the controlling of the sound intensity and audio dynamic range of the incoming analog audio signal with the digital audio recording software and hardware, it is suggested that these engineers should choose the session parameter setting to 24-bit (Huang, 2020).

(3) As for the choices of sample rate, the experimental results showed that there was not much difference in the respondents' satisfaction level between the session settings at 44.1 and 48 kHz, with the psychological evaluations in audio dynamic range.

(3) In addition, based on the Nyquist theorem, the session setting of 44.1 kHz in digital audio recording system is capable and enough to record audio frequency up to 22.05 kHz, which is higher than the upper range of human hearing (20 kHz). According to the aforementioned statements, there is not much difference in auditory satisfaction between 44.1 and 48 kHz. Thus, for the consideration of computer disk storage consumption and accommodation, the session setting at 44.1 kHz is the best recommendation of this research.

REFERENCES

Aji, B. M., & Larner, A. J. (2017). Screening for dementia: single yes/no question or Likert scale? *Clinical Medicine Journal, 17*(1), 93–94. doi: 10.7861/clinmedicine.17-1-93

Cook, F. D. (2017). *Pro Tools 101: Pro Tools Fundamentals I - Version 12.8.* Avid Technology, Inc.

Grondin, F., Tang, H., & Glass, J. (2020). Audio-visual calibration with polynomial regression for 2-D projection using SVD-PHAT. Paper presented at the *ICASSP 2020- 2020 IEEE International Conference on Acoustics, Speech and Signal Processing (ICASSP),* pp. 4856–4860, doi: 10.1109/ICASSP40776.2020.9054690

Hennink, M. M., Kaiser, B. N., & Weber, M. B. (2019). What influences saturation? estimating sample sizes in focus group research. *Qualitative Health Research, 29*(10), 1483–1496. https://doi.org/10.1177/1049732318821692

Huang, C.-C. (2020, May). Discussions on technical specifications of digital audio sampling. Paper presented at the *International Conference on Innovation Digital Design 2020,* Tainan, Taiwan.

Isnawati, A. F., Citra, V. O., & Hendry. J. (2019). Performance analysis of audio data transmission on FBMC-Offset QAM system. Paper presented at the *2019 IEEE International Conference on Industry 4.0, Artificial Intelligence, and Communications Technology (IAICT),* pp. 81–86. doi: 10.1109/ICIAICT.2019.8784810

Mao, Y., & Monahan, A. (2017). Predictive anisotropy of surface winds by linear statistical prediction. *Journal of Climate, 30*(16), 6183–6201. https://doi.org/10.1175/JCLID-16-0507.1

Mishra, P., Pandey, C. M., Singh, U., Gupta, A., Sahu, C., & Keshri, A. (2019). Descriptive statistics and normality tests for statistical data. *Annals of Cardiac Anaesthesia, 22*(1), 67–72. https://doi.org/10.4103/aca.ACA_157_18

System Innovation in a Post-Pandemic World – Kin-Tak Lam et al. (Eds)
© 2022 Copyright the Author(s), ISBN: 978-1-032-24392-4

Research on the reproducibility of the application of projection mapping on restaurant service

Wen-Yuh Jywe, Liang-Yin Kuo, Siu-Tsen Shen & Tzu-Yi Chen
National Formosa University, Huwei Township, Yunlin, Taiwan, ROC

ABSTRACT: The purpose of this study is to bring Projection Mapping into the restaurant service to change the traditional monotonous restaurant service. The use of Projection Mapping makes each process full of surprises, adding to the dining atmosphere. During the early stage of the research, first, we collected data by Document Analysis and organized Projection Mapping cases, and conducted expert interviews as research, and finally, we presented the creation with store cases. In the creation of the projection, the author used the projection of the service in the case of meals, changing the traditional way of the staff introducing the menu, and using the projection of the whole space and the table projection to convey the content, so that customers can immerse themselves in the dining environment. The Projection Mapping sets up a projector device, "Lightform", and uses a "LFC" device to scan the scene and project the tables and environment in alignment and connects to the desktop software "Lightform Creator" to create the images. The technology of Projection Mapping makes dining more interesting than a single purpose, enhances the sense of anticipation of dining and improves the service quality of the restaurant, and achieves a new media art experience at the same time. As the form of art is not limited, this study also hopes to serve as a reference for the future use of Projection Mapping in the restaurant industry for commercial purposes.

Keywords: Immersive experience, Projection Mapping, Restaurant Service

1 INTRODUCTION

In recent years, projection mapping is widely used in building projection, landscape projection and various artistic creations and outdoor performances. With more varied visual effects, it can at- tract the attention of the masses more effectively.

The expansion of this technology is not limited to visual experience, but also a combined immersive concept. Participants can interact with the work and blend into the environment to achieve the effect of immersing the five senses into the work.

Nuit Blanche is held in many countries, including Taiwan, and is an art exhibition that empsizes the integration of art with the environment. It conveys the beauty of the city through light interac- tive installations such as large-scale installation art and projection mapping shows.

This research mainly discusses the commercial viability of projection mapping applied to catering services. For this reason, it is in collaboration with Yunlin's sweet potato and pastry specialty store "HU JHEN TANG" (虎珍堂) in the store Projection design of food inside. Focusing on the store history of "HU JHEN TANG" and the local ingredients in Yunlin, customers can learn about the food production process and historical appearance while enjoying meals (Figures 1–4).

Figure 1. 2017 Taipei Nuit Blanch (extracted from Department of Cultural Affairs, Taipei City Government).

Figure 2. HU JHEN TANG's food photo (extracted from HU JHEN TANG).

2 EXPERIMENTAL SECTION

The purpose of this research is mainly to attract customers and meet the two needs of store marketing. During the experience, the use of music and table projections can bring customers a sense of freshness from the past, and they can learn about the products sold by the store and the store story through the projected screen during the meal.

First of all, this study through the Interview method with the boss - Mr. Lin Jincheng learned after the interview, "HU JHEN TANG" historical origins of this shop.

In 1960, the shop was formerly known as Zhengyi Department Store, and it was located on the first street of Huwei Town, Yunlin (now Zhongshan Road, Huwei Town, Yunlin), which was the busiest street in the town at that time; for this reason, the owner kept the store design. The exterior and internal structure of the Japanese-style building of Zhengyi Department Store at that time were unchanged. The furniture in the store was decorated with old antiques, and some related historical stories were written on the walls, so that customers who came to dine could also eat and at the same time feel such a retro atmosphere. In terms of meals, "HU JHEN TANG" mainly sells sweet potato-related products, including Han Ji Nong cheesecake, sweet potato crisps, sweet potato nougat and so on.

Figure 3. Old photos of Zhongshan Road in the 1960s (extracted from HU JHEN TANG).

The dining items in the store are mainly "hu yue shao"(虎月燒), "hu yue shao" is honey dorayaki pie crust filled with sweet potato cheese, and adzuki beans, including matcha, purple sweet potato, sweet potato Three flavors are available. Other products are displayed on the first floor. The research will be divided into three stages: The first stage is to make animation graphics in multimedia editing software - After Effects. In the second stage, the Lightform device LFC is used to set up the area to be projected and the environment. In the third stage, the film and animation are output, and the effect is projected on the dining table.

Figure 4. "hu yue shao" (extracted from HU JHEN TANG).

3 RESULTS AND DISCUSSION

3.1 *The first stage: Assessment of the environment, animation settings*

For the projection mapping of "HU JHEN TANG"'s catering service, first a site survey was conducted to understand the original dining environment of the guests and analyze the advantages and disadvantages of the environment. After discussing with the boss after the site survey, I decided to focus on the historical evolution, introduction of shops and meals, and animation projections during meals.

First of all, in the part of presenting history and culture, old photos are used to show the appearance of the time, and a short text description describes the difference between the past and the present (Figures 5–9).

Figure 5. Set picture animation in after effects.

The design of eating "hu yue shao" includes the projection of product introduction in the store, the purpose is to achieve the purpose of marketing other products by watching the projection while dining.

Figure 6. Designed the introduction screen of HU JHEN TANG products in illustrator.

3.2 Second stage: Lightform create settings

The creative software-Lightform Create that integrates projection settings and positioning and can simply design the projection range. With the Lightform Compute (LFC) kit and camera, after scanning the range to be projected, it can be used with projectors of various focal lengths without being restricted by the environment. It greatly simplifies the pre- projection work; in addition, the built-in mask tool of Lightform Create can use anchor points or magic wands to select the range, which is very convenient. And the built-in animation effect and layer style can make the projection more variability.

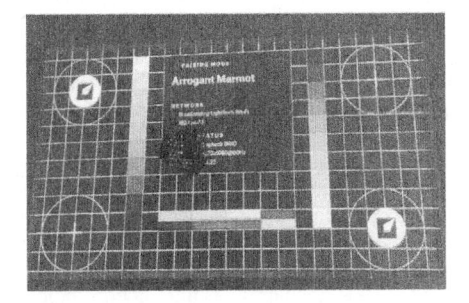

Figure 7. LFC projection setting screen.

3.3 The third stage: Projection during meal

It is made with the theme of "hu yue shao" sold by HU JHEN TANG. The original "hu yue shao" is an afternoon snack that can be eaten while walking. However, this effect cannot be achieved when dining in the store, so the projection is used. The mapping technique presents the artistic con- ception of roaming Huwei when dining in the store, and the scene of Huwei is projected through the screen to create two feelings of enjoying the local scenery while dining in the store.

Figure 8. Projecting the landscape on the dining table.

Figure 9. Projecting the landscape on the dining table.

In addition, in other food display of "HU ZHEN TANG", changes are made on the packaging box through animation projection to attract the attention of customers.

Figure 10. Projection on the packaging box.

4 CONCLUSION

Most of the more common projection creations in Taiwan are outdoor large-scale projection shows, and few cases of table projection are found. Foreign restaurants have more similar creative methods.

Projection mapping is a method of attracting customers. Although it can make customers feel fresh and integrate into this kind of projection environment, it lacks interactive devices at present. It can only be experienced by hearing and vision. If you can add light sculpture projection Interactivity, simple game quiz levels, can enhance the fun of dining.

This time, Lightform is mainly used for projection design, but the disadvantage of Lightform is that in addition to the built-in animation, if you want to make more complete story animation content, you must rely on AfterEffect or Premiere to assist in the design. But from the point of view of food service, it is no longer a difficult thing to create a projection on the storefront. The interface design of Lightform Create is simple, with few functions and precise; there is no specification limit on the choice of projector and projection. The setting is no longer limited to one location. As long as the creator wants to project the location or object, Lightform Compute (LFC) will be able to figure out any objects that are not flat.

This was an example of a projection mapping for a food service. I hope that this research will give more marketing ideas to more businesses who want to try innovation. In addition, it can also give creators new creative thinking in using Lightform.

REFERENCES

Wikipedia- nuit blanche. Retrieved May 18, 2021 from https://reurl.cc/lbn8jx. Wikipedia- projection mapping. Retrieved May 20, 2021 from https://reurl.cc/lbn839 lightform. Retrieved May 25, 2021 from https://lightform. com/

System Innovation in a Post-Pandemic World – Kin-Tak Lam et al. (Eds)
© 2022 Copyright the Author(s), ISBN: 978-1-032-24392-4

A framework of innovative product development process based on expected final result

Tien-Lun Liu, Hung-Huai Shen, & Yi-Hsuan Chen
Department of Industrial and Systems Engineering, Chung Yuan Christian University, Chung-Li, Taoyuan, Taiwan

ABSTRACT: If a new product cannot be developed proficiently, it will be replaced soon by competitors' products. Therefore, efficiency and effectiveness are very important for product development. This research proposes to establish an innovative product development process model that is inspired by the "Ideal Final Result" in the TRIZ theory. The product design process will be carried out starting from the "Best Final State." First of all, we defined the final state that the product design which will be satisfied with all the requirements as "Expected Final Result (EFR)." Such EFR might face conflicts among design requirements. However, if the EFR cannot be achieved, the development team must consider the next goal, which is called "the Second Best State." Then how to define the secondary ideal result needs a systematic analytic mechanism, rather than just directly deleting certain design requirements. We combined the "four-level data structure" to link design requirements as well as assist in establishing the relationships among products and TRIZ parameters. In this framework, we set three indicators that help define the secondary EFR, including "Contradiction indicator," "Demand-based indicator," and "Demand complexity indicators." By comparing the correlation between current product design and the parameters, we determine the condition to obtain the next design goal using the indices. By using different indicators, we may reach different options for product design. According to the above-mentioned decision-making mechanism, a product development process can be constructed to achieve the goal of "starting from the Ideal Final Result" for continuous improvement, which is beneficial to achieve a distinct innovation.

1 INTRODUCTION

In order to make products more competitive, "efficiently and continuously improvement" has become the primary goal of every product development team. Problem identification and solving have to be processed with limited resources in the development projects that need to achieve more effective results. Collaborative product development has been the demonstration of the implementation of concurrent engineering in enterprises, the efficiency of "divide and conquer" will be the key to fulfill product design requirements (DRs), quality, and performance.

To achieve innovation and efficiency of product development by resolving contradictions within the process, this study will explore the product development model by "starting from the Ideal Final Result" and establish the second-best indicator to improve the efficiency of resolving conflicts while they occur. Combined with the basic concept of TRIZ theory, we may realize product design and development models that can effectively improve innovation. When there exist contradictions that are difficult to overcome, we should apply certain comprehensive indicators to assist the development team to find options suitable for the product development direction, and then make the selection and redefine the goal for the next

stage. In such an approach that "starting from the Ideal Final Result" for continuous improvement can be constructed.

2 LITERATURE REVIEW

2.1 *TRIZ*

TRIZ theory originated from four basic concepts, namely "Functionality," "Use of Resources," "Contradictions," and "Ideality." Based on these concepts, various methods and tools have been developed, resulting in a set of methods that can adapt to different situations and find trigger solutions to different problems. The contents include: Problem Formulation, Functional Analysis, Contradiction Matrix, 40 Inventive Principles, Trends, Substance-Field Analysis, Ideal Final Result (IFR), Effects, and TRIZ algorithm (ARIZ). This theory can help obtain ideas for solving problems in different areas, it can also stimulate thinking about difficulties encountered in product development through systematic analysis. However, the assistance of TRIZ theory can break the knowledge limitation and provide a systematic solution to the problem.

At first, the TRIZ originator Altshuller established the corresponding relationship between 39 engineering parameters (EPs) and sorted them into

DOI 10.1201/9781003278474-4

a 39 × 39 contradiction matrix. Users can find out the invention principle that can resolve the contradiction from the contradiction matrix by using two EPs generated by the system. D. Mann proposed Matrix 2003 and extended the EPs to 48 items as shown in Table 1.

Table 1. Forty-eight engineering parameters.

1.	Weight of moving object
2.	Weight of stationary object
3.	Length/angle of moving object
4.	Length/angle of stationary object
5.	Area of moving object
6.	Area of stationary object
7.	Volume of moving object
8.	Volume of stationary object
9.	Shape
10.	Amount of substance
11.	Amount of information
12.	Duration of action of moving object
13.	Duration of action of stationary object
14.	Speed
15.	Force/torque
16.	Energy used by moving object
17.	Energy used by stationary object
18.	Power
19.	Stress/pressure
20.	Strength
21.	Stability
22.	Temperature
23.	Illumination intensity
24.	Function efficiency
25.	Loss of substance
26.	Loss of time
27.	Loss of energy
28.	Loss of information
29.	Noise
30.	Harmful emissions
31.	Other harmful effects generated by system
32.	Adaptability/versatility
33.	Compatibility/connectability
34.	Ease of operation
35.	Reliability/robustness
36.	Repairability
37.	Security
38.	Safety/vulnerability
39.	Aesthetics/appearance
40.	Other harmful effects acting on the system
41.	Manufacturability
42.	Manufacture precision/consistency
43.	Automation
44.	Productivity
45.	System complexity
46.	Control complexity
47.	Ability to detect/measure
48.	Measurement precision

2.2 Expected final result

The Expected Final Result (EFR) is defined to solve the conflict caused by the collaborative development of the product, and the state of the product is to meet all the DRs of the collaborative development team, such concept is different from the original definition of "Ideal Final Result (IFR)" in TRIZ, which is described as "when the system evolves to achieve all the required benefits (functions) and none of the harmful things." Nevertheless, the state of IFR is not easy to realize and accomplish.

After proposing the definition of the EFR model from the perspective of product collaborative development, there is a relatively clear goal and direction. After defining the most perfect product design result state, it will be improved by the following process, as shown in Figure 1, assuming that the ultimate goal to be achieved is EFR(0), which means that all DRs and conditions must be met. By comparing the current product status with the EFR(0) converted from all the DRs of the development team, and using the evolutionary line in TRIZ's trends of evolution to analyze, we may grasp the current product status and EFR(0), explore the evolving reasons of these evolution trends, and then establish a correlation with TRIZ's 48 EPs. After the correlation with the EPs is established, it can be converted into a basis for the collaborative development team to resolve contradictions in practice. If there is a contradiction that cannot be solved, the product state EFR(0) must be modified to EFR(1), and then the technical evaluation is carried out, etc., until a feasible design is formed.

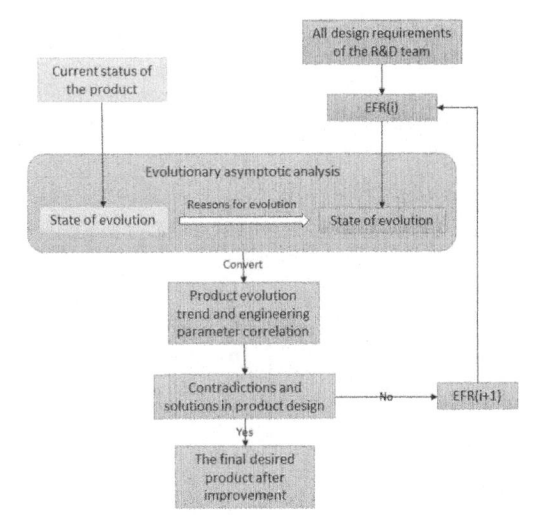

Figure 1. The final expected result development process architecture diagram.

2.3 Four-level data structure

Since the requirements and conditions of product design come from the team members of collaborative design, each of which has its own consideration and purpose, a four-level data Structure classification model is proposed to unify the DRs and conditions of product design, and to assist in establishing the correlation of TRIZ parameters. The description of each level is as follows:

1. The first level is "Design Intention (DI)." Use the concept of Design for X (DFX) to classify every design condition to realize the DIs of the co-design members. This "X" may include purposes such as customer requirements, manufacturability, environmental impact, etc.
2. The second level is "Design Requirement (DR)." It is the design request from the co-developers for the product, and links to the design purpose to indicate the reason for this requirement.
3. The third level is "Design Parameter (DP)." It can be described as the design specifications or components of the product, and it is crucial for team members to examine the contradictions within the design.
4. The fourth level is "Engineering Parameters (EP)," which refers to the 48 EPs defined by Matrix2003. It helps catch the key characteristics of the DPs, as well as the essence of product design restrictions.

This four-level relational model classifies product design conditions in the conceptual design stage to establish a structured information processing framework and subsequent product analysis basis to link to the problem-solving methods in TRIZ. Figure 2 is a schematic diagram of the concept.

After the design purpose, DRs, DPs, and EPs are defined according to the classification conditions of the four-level relation, relevant matrices can be established according to their correlations. At the beginning, the correlation of each level can be expressed by constructing a forward correlation matrix as follows:

$$\{DI\}_{p \times 1} = [A]_{p \times n}\{DR\}_{n \times 1} \tag{1}$$

$$\{DR\}_{n \times 1} = [B]_{n \times m}\{DP\}_{m \times 1} \tag{2}$$

$$\{DP\}_{m \times 1} = [C]_{m \times r}\{EP\}_{r \times 1} \tag{3}$$

Among them, DI is the design purpose, with a total of p items; DR is the design requirement, with a total of n items; DP is the design parameter, with a total of m items; EP is the TRIZ EP, with a total of r items, and A, B, and C are used to represent the association matrix of each class. For example, the corresponding relationship between DRs and DPs is expressed by an association matrix as follows:

$$
\begin{array}{ccc}
\text{DR} & \text{AM} & \text{DP}
\end{array}
$$

$$
\begin{Bmatrix} DR_1 \\ \vdots \\ \vdots \\ DR_n \end{Bmatrix} =
\begin{bmatrix}
B_{1.1} & B_{1.2} & \cdots & B_{1.m} \\
B_{2.1} & B_{2.2} & \cdots & \vdots \\
\vdots & \vdots & \ddots & B_{n-1.m} \\
B_{n.1} & \cdots & \cdots & B_{n.m}
\end{bmatrix}
\begin{Bmatrix} DP_1 \\ \vdots \\ \vdots \\ DP_m \end{Bmatrix} \tag{4}
$$

$$
B_{nm} = \begin{cases} 1 & \text{When } DR_i \text{ is related to } DP_j \\ 0 & \text{When } DR_i \text{ is independent of the } DP_j \end{cases}
$$
$$
i = 1 \ldots n, \qquad j = 1 \ldots m
$$

In the following section, we will use the correlation scheme of the four-level structure to establish the relevant equation and apply it to the quantitative decision model, and to realize and discuss the current situation of the product design problem.

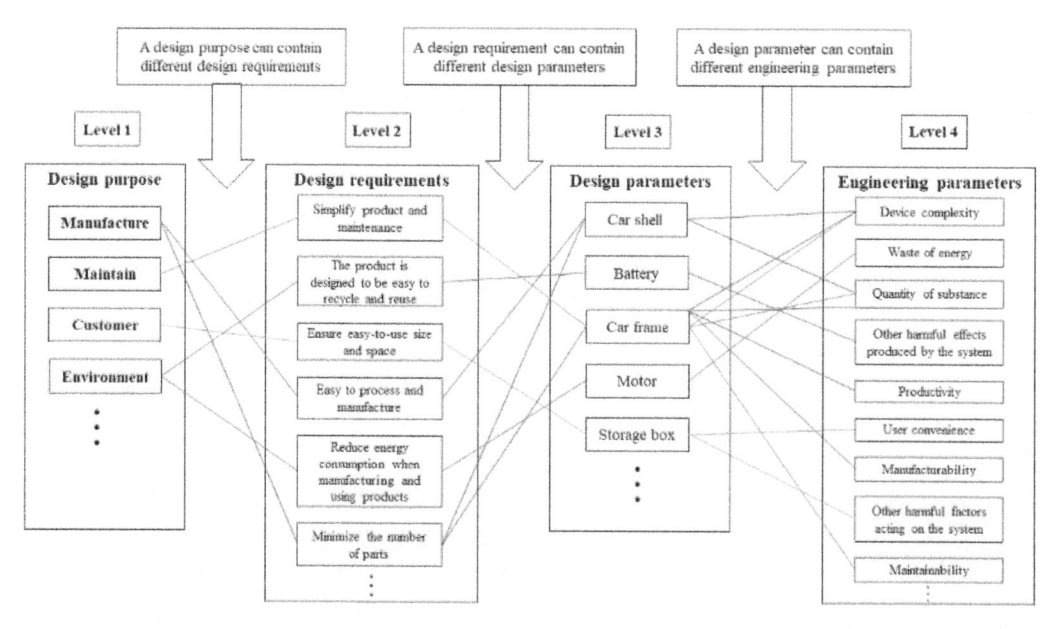

Figure 2. Illustration of four-level data correlation structure.

3 METHODOLOGY

3.1 EFR-oriented product development process

The product development process architecture proposed by this research is shown in Figure 3, which will be based on the product development process model of EFR. Since the process of this research focuses on the relationship between DPs and EPs, the part of the evolutionary state analysis in the evolutionary asymptote is simplified, and the concept of four-level correlation is added to modify the process.

In the beginning, the DRs are defined as EFR(0), where EFR(0) represents the optimal situation and means that all the current DRs are met. Then, the information and design conditions of the product will be converted and classified in the form of a four-level framework. After the conversion, the contradictions in the product design will be analyzed and resolved. If it cannot be solved, EFR(1) will be redefined and continuously improved until all DRs can be met.

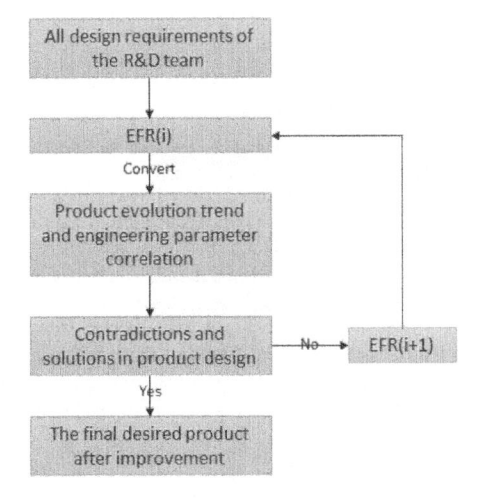

Figure 3. Product development process architecture diagram.

3.2 Integration and analysis of contradictions

After the final expected result EFR(0) is defined, we will start to explore whether all the DRs can be met. When the needs cannot be fully met, we must find out where the conflicts are and eliminate them. The contradiction in DRs is the priority to solve the contradiction, but the contradiction may occur in multiple places, and between different sets of parameters. Through the four-level relational structure, it helps to clarify the logic of contradiction analysis and the order of handling problems. Use the relationship between DRs, DPs, and EPs in the associative architecture to establish the relevance and influence of various requirements, and analyze whether the contradiction can be gradually solved by this approach. If the contradiction cannot be resolved, the quantitative mechanism for deciding "the second best" situation will be applied to finding EFR(1) for the next possible product improvement.

3.3 EFR quantitative mechanism to assist decision making

In this section, we will discuss how to establish the processing indicators of "the second best." These indicators will affect the precedence of conflict resolution. Therefore, before defining the second-best indicators, we will formulate processing rules from different aspects, such as "Contradiction indicator," "Demand-based indicator," and "Demand complexity indicators." The following three processing rules will be explained:

1. Contradiction indicator: The contradiction indicator refers to the number of EPs or DPs involved, which may imply the complexity of the contradiction.
2. Demand-based indicator: It is related to the importance of DRs. The so-called importance refers to determining the weight setting when defining the DRs.
3. Demand complexity indicators: Based on the mixed consideration of the above rules. When the contradiction occurs, it may be faced with the same number of EPs or DRs corresponding to the design requests. In this case, the weight of DRs will be used to make a judgment.

These three processing principles and concepts will be discussed as follows:

3.3.1 Contradiction indicators

The first indicator is a contradiction indicator. If the number of EPs and DPs involved in the contradiction is the largest, priority will be given to the EPs and whether the DPs can be removed from the DRs. When one of the DPs in the contradiction can be removed, the contradiction can be resolved by this, otherwise, the corresponding DR needs to be removed.

Before establishing the relationship equation, the relationship between the various classes will be clarified by the forward incidence matrix. In the case of conflicts, the relationship between conflicts and DPs will be discussed first, as shown in (5). Assuming that a total of p contradictions and m DPs occurred during product development, CCq is used to represent the qth contradiction and conflict, and DPk is used to represent the kth DP. The correlation matrix can be used to understand the correlation between the contradiction conflict and the design parameter. Dqk is used to express the correlation between the qth contradiction conflict and the kth DP.

$$
\begin{matrix} CC \\ \begin{Bmatrix} CC_1 \\ \vdots \\ \vdots \\ CC_p \end{Bmatrix} \end{matrix} = \begin{matrix} CM \\ \begin{bmatrix} D_{1.1} & D_{1.2} & \cdots & D_{1.m} \\ D_{2.1} & D_{2.2} & \cdots & \vdots \\ \vdots & \vdots & \ddots & D_{p-1.m} \\ D_{p.1} & \cdots & \cdots & D_{p.m} \end{bmatrix} \end{matrix} \begin{matrix} DP \\ \begin{Bmatrix} DP_1 \\ \vdots \\ \vdots \\ DP_m \end{Bmatrix} \end{matrix} \quad (5)
$$

$$
D_{qk} = \begin{cases} 1 \text{ When } CC_q \text{ is related to } DP_k \\ 0 \text{ When } CC_q \text{ is independent of the } DP_k \end{cases}
$$

$$q = 1 \ldots p, \qquad k = 1 \ldots m$$

After understanding the relationship between conflicts and DPs through the above correlations, we can establish a mathematical relationship between the two, as shown in (6). The contradiction indicators value is obtained by using the sum of the associated values of the DPs corresponding to the contradictions.

$$CC_q = \sum_{k=1}^{m} D_{q.k} \tag{6}$$

Using the above formula, the sum of the number of DPs involved in the contradiction in item q can be obtained. The one with the largest number will be selected and evaluated whether the relevance to the corresponding DRs can be removed. If the DRs cannot be removed, EFR(1) can be generated.

3.3.2 Demand-based indicator

The second index is the demand index. The purpose of this index is to complete all the requirements, so as to achieve the highest satisfaction degree of the requirements. If all the requirements cannot be completed, the one with the lowest weight of the DRs will be found and removed, and the following relationship will be constructed on this basis, as shown in (7). It is assumed that all the demands have n items, and the judgment is made by summing up the percentage of demand weights, so as to find the combination of the highest demand satisfaction degree. In the formula, Max DR represents the degree of satisfaction of the DRs, DRi represents the ith DR, and dri is used to express whether the ith DR is met. Under the ideal condition when all DRs are met, this indicator indicates that all dri are 1, that is, the highest demand satisfaction is 100%. However, in reality, there may be conflicts between requirements, so all requirements cannot be met. In the case of abandoning the requirement, dr_i can be 0, and e_i represents the weight percentage of the ith DR. This weight comes from the consideration and discussion of each DR by the members of the development team, if the development team conducts QFD.

$$MAXDR = \sum_{n=1}^{n} e_i * dr_i = e_1 dr_1 + e_2$$
$$dr_2 + \cdots + e_i dr_i$$
$e_i, i = 1, \ldots, n$ The weight percentage of the DR_i
$dr_i, i = 1, \ldots, n$ Indicates whether the DR_i is met
$dr_i = 1$, Indicates that the DR_i is met
$dr_i = 0$, Indicates that the DR_i is not met

$$\tag{7}$$

The above equation is a thinking model based on DRs. The following will propose an equation combining DRs, DPs, and EPs.

By exploring the relationship between the three, we will provide the most effective way to assist development teams in resolving contradictions in different situations.

3.3.3 Demand complexity indicators

The last indicator is the demand complexity indicator. This indicator only discusses the relationship between DRs themselves and DPs and EPs. The relationship between the DRs and the DPs will be defined first. If the DR involves the most DPs, in order to reduce the chance of conflict with other requirements, it will be deleted first and then proceed to the next best.

We use the association matrix to discuss the relationship between DRs and DPs, as shown in (8). Assuming that a total of n DRs and m DPs occurred during product development, DRi is used to represent the ith DRs, and DPk is used to represent the kth DP. The correlation matrix can be used to understand the correlation between the DRs and the DP. Bik is used to express the correlation between the ith DRs and the kth DP.

$$
\begin{matrix} DR & CM & DP \end{matrix}
$$

$$
\begin{Bmatrix} DR_1 \\ \vdots \\ \vdots \\ DR_n \end{Bmatrix} = \begin{bmatrix} B_{1.1} & B_{1.2} & \cdots & B_{1.m} \\ B_{2.1} & B_{2.2} & \cdots & \vdots \\ \vdots & \vdots & \ddots & B_{n-1.m} \\ B_{n.1} & \cdots & \cdots & B_{n.m} \end{bmatrix} \begin{Bmatrix} DP_1 \\ \vdots \\ \vdots \\ DP_m \end{Bmatrix} \tag{8}
$$

$$B_{i.k} = \begin{cases} 1 \text{ When the } DR_i \text{ is related to the } DP_k \\ 0 \text{ When the } DR_i \text{ is independent of the } DP_k \end{cases}$$
$q = 1 \ldots p, \qquad\qquad k = 1 \ldots m$

After understanding the relationship between DRs and DPs through the above correlations, we can establish a mathematical relationship between the two, as shown in (9). The sum of the associated values of the DPs corresponding to the DRs is used to obtain the number of DPs involved in the DRs, that is, the demanding complexity indicators value. From this formula, the sum of the number of DPs involved in the ith demand can be obtained. The one with the largest number involved will remove its design demand, and EFR(1) can be generated.

$$dr_i = \sum_{k=1}^{m} B_{i.k} \tag{9}$$

The judgment of the requirement complexity index will use the correlation between DRs and DPs as the main judgment tool. There are two possible situations in which the above method cannot be used to perform this indicator: the first case is that when we are developing a new product, there are no components yet, so the DPs are not easy to identify. In the second case, the product needs to maintain the existing functions, but when the components need to be replaced, the DPs will change accordingly.

In these cases, we will use the EPs corresponding to the DRs to perform the demand complexity indicators. First of all, we also use the correlation matrix to observe the relationship between DRs and EPs. As shown in (10), assuming that there are n DRs and r EPs during product development, and there are only 48 EPs in total, so r ≤ 48, DRi is used to represent the DR of item i, EPj is used to represent the EP of item j. The correlation matrix can be used to understand the correlation between DRs and EPs. Bi.k is used to represent the correlation between the DRs of item i and the DPs of item k.

$$\begin{Bmatrix} DR_1 \\ \vdots \\ \vdots \\ DR_n \end{Bmatrix} = \begin{bmatrix} B'_{1.1} & B'_{1.2} & \cdots & B'_{1.m} \\ & B'_{2.1} & B'_{2.2} & \cdots & \vdots \\ \vdots & \vdots & \vdots & \ddots & B'_{n-1.m} \\ B'_{n.1} & \cdots & \cdots & B'_{n.m} \end{bmatrix} \begin{Bmatrix} EP_1 \\ \vdots \\ \vdots \\ EP_r \end{Bmatrix} \tag{10}$$

$$B'_{i,j} = \begin{cases} 1 \text{ When the } DR_i \text{ is related to the } EP_j \\ 0 \text{ When the } DR_i \text{ is independent of} \\ \quad \text{the } EP_k \end{cases}$$

$$i = 1 \ldots n, \qquad j = 1 \ldots r, r \leq 48$$

After understanding the correlation between DRs and EPs through the above correlation formula, the mathematical relationship between the two will be established, as shown in (11). The sum of the associated values of the EPs corresponding to the DRs is used to obtain the number of EPs involved in the DRs, so as to replace the alternative scheme when the DPs cannot be identified. From this formula, the sum of the number of DPs involved in the ith demand can be obtained. The one with the largest number involved will remove its design demand, and EFR(1) can be generated.

$$dr_i = \sum_{v=1}^{r} B'_{i,j} v = 1 \tag{11}$$

4   CONCLUSIONS

This paper proposes an EFR concept from IFR with TRIZ theory that constructing into a product development process framework. We apply the four-class associative model for design condition classification to find out the design purpose, DPs and EPs

corresponding to the DRs. Under the four-class data architecture, it helps contribute to the logic of contradiction analysis and the order of problem handling. Once there is an unresolvable contradiction, the three indicators of "the Second Best" will be applied to help determine EFR(1), and then redefine the design goal. The three indicators are "Contradiction indicator," "Demand-based indicator," and "Demand complexity indicators," which can provide different product development directions and assist the development team to explore the next step. According to the above-mentioned decision-making mechanism, a product development process can be constructed to achieve the goal of "starting from the Ideal Final Result" for continuous improvement, which is beneficial to achieve a distinct innovation.

REFERENCES

Kai-Jen Cheng, A Systematic Analysis Model for Product Design and Quality Traceability, Department of Industrial and Systems Engineering, Chung Yuan Christian University College of Management Graduate Thesis, 2019

Mann, D. Hand-on Systematic Innovation, Belgium: CREAX Press, 2002.

Mann, D., Zlotin, B. Zusman, A. Matrix 2003: Updating the TRIZ Contradiction Matrix, CREAX, Press, Detroit, USA, 2003.

Tien-Lun Liu, Based on TRIZ to Combine AI Technology and Quantitative Mechanism to Establish Expected-Final-Result Oriented Model for Collaborative Product Development, Ministry of Science and Technology Project, 2018. (MOST 108-2221-E-033 - 012 -MY3)

Wei-Sheng Huang, IMPD Model in Collaborative Product Development Process, Department of Design, National Taiwan University of Science and Technology Graduate Thesis, 2013

Yi-Chen Li, A Framework of Hierarchical Representation for Product Design Request, Department of Industrial Management and Business Administration, St. John's University Graduate Thesis, 2015

System Innovation in a Post-Pandemic World – Kin-Tak Lam et al. (Eds)
© 2022 Copyright the Author(s), ISBN: 978-1-032-24392-4

Attention module combined with A-U-Net architecture for image matting

Guodong Zhang & Shwu-Huey Yen

Department of Computer Science and Information Engineering, Tamkang University, New Taipei, Taiwan

ABSTRACT: Trimap plays an essential role in image matting. However, generating trimaps is expensive. Instead, we propose an architecture named A-U-Net equipped with both spatial and channel attention modules to estimate alpha matte without trimaps. However, there are two opposite phenomena observed in estimating alpha matte. The fine-structure edges connecting to the background should be preserved, whereas the texture details inside the foreground should be ignored. Thus, alpha matte is prone to making errors when the foreground is cluttered with texture or when both the foreground and the background have similar textures. To alleviate this problem, we design an additional classifier to learn the pure foreground from the input image. This "pure" indicates that those pixels must be foreground ones. Then, we add it to the original A-U-Net architecture result and successfully solve the problem. Extensive experiments have proved that our method is comparable or even better than other state-of-the-art methods.

1 INTRODUCTION

Image matting is the process of separating a foreground object from an image, which is a common practice in image and film editing. Especially with the development of virtual reality and augmented reality technologies, image matting is increasingly gaining attention in the image editing field. However, matting techniques are not able to keep up with the current demand due to the difficulties involved in solving the problem discussed below. Mathematically, image I is considered as a linear combination of foreground image F and background image B based on the alpha matte α such that at each pixel i

$$I_i = \alpha_i F_i + (1 - \alpha_i) B_i, \alpha_i \in [0, 1], \tag{1}$$

where $\alpha_i \in [0, 1]$ is the alpha value at pixel i Eq. (1) can be interpreted as follows: given an input image I, to separate F from the image, F, B, and α are estimated simultaneously. This is an ill-posed problem for there are seven unknowns but only three values are known in the formula. To alleviate this matting problem, previous approaches usually impose a trimap to constrain the alpha matte from the input image [1–7]. However, the quality of alpha matte is related to the provided trimap and creating a good trimap is expensive and difficult for general users.

The introduction of the attention module in the image recognition domain is a very crucial technique. In the RA-CNN model [8], the attention network is concerned not only with the overall information but also with the local information. Inspired by this, for image matting, the attention module is used to guide the network to pay more attention to the foreground and less attention to the background. This enables

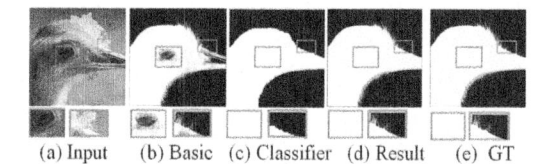

| (a) Input | (b) Basic | (c) Classifier | (d) Result | (e) GT |

Figure 1. The output of different stages of the proposed network: Basic is the output of A-U-Net module with the attention block Classifier is the prediction of pure foreground (details in Section 3.2) Result is obtained from (b) + (c).

image matting to focus on the image foreground without a clue of trimap [9, 10].

There is matting research ongoing with input images as the only information. HattMatting [9] employs a spatial-wise attention module to extract appearance cues in the low stage and a channel-wise attention module to focus on semantic features in the high stage Late Fusion [11] utilizes two decoder branches for foreground/background classification and fuses the classification results to obtain the final alpha values through a fusion network. These approaches usually do not produce satisfactory results when a cluttered foreground is encountered or when both the foreground and the background have similar colors or textures.

As shown in Fig. 1, when a trimap is not used, (b) is the result of A-U-Net with an attention block, which is very effective at the fine structure of the border, but the texture details of the foreground interior will also be displayed. The pure white part of the classification result can remove the foreground interior details as shown in Fig. 1(c). With the help of a predicted opacity

DOI 10.1201/9781003278474-5

area, we can obtain a very good result as shown in Fig. 1(d).

We observe that when the foreground has fine textures, such as fine hairs, the alpha matte needs to elaborate those fine edges. But when the foreground has a texture on itself we do not want the network to catch those textures. These two scenarios are quite opposite. This inspired us to make these two cases consistent instead of contradicting each other. Therefore, we add a classifier for pure foreground recognition. In a pure foreground, the system only needs to detect a rough contour without fine details.

2 RELATED WORK

2.1 Image matting

This section is divided into two parts: the matting method with a trimap as an input and the matting method without a trimap as an input.

Most previous methods, including traditional methods and early machine learning methods [1–7] use a trimap as an additional input Traditional methods [1, 2] are based on low-level color or structural features that are not as recognizable in opaque areas or fine edges. DCNN [3] claims to be the first to use deep learning for natural image matting. Deep image matting [4] integrates RGB images and trimaps, and it uses high-order semantics to estimate alpha mattes. An adobe matting dataset was also proposed to provide a dataset for the subsequent deep learning methods to estimate alpha mattes. After that, many methods [5–7] for image matting based on machine learning were proposed. Although they have significant positive, they are limited by the use of trimaps as an additional input, which makes practical applications expensive

Since a trimap is very expensive, many methods have abandoned it in recent years, which is beneficial to practical applications [8–10] In semantic human matting [8], a trimap is not used as an additional input for the first time; instead, it uses an additional framework to estimate the mixed area, and then uses the main framework to refine it. A Glance and Focus Matting (GFM) network in animal image matting [10] employs a shared encoder and two separate decoders to learn both tasks in a collaborative manner for end-to-end animal image matting.

2.2 Attention module

The attention mechanism has been applied to the image field since long. In [17] the Google Brain team uses an attention mechanism in the RNN model for image classification and achieves a very good performance Since then, the attention mechanism has been widely used in the NLP and image processing Many novel feature expression modules based on the attention mechanism have been proposed. Convolutional Block Attention Module (CBAM) [11] extends the attention map along two separate dimensions of feature maps: channel and spatial They utilize both max-pooling and average-pooling to capture distinct and all feature information. In [13], a novel trainable Second-Order Channel Attention (SOCA) module has been developed for achieving a more discriminative representation by using second-order statistics of channel-wise features. One noteworthy characteristic of these attention modules is they can be integrated with any CNN architecture seamlessly with negligible overheads.

3 OUR APPROACH

The matting problem can be solved in two steps: The first step is to identify the semantic foreground of the image and the second step is to refine the foreground edges. To avoid the need of trimaps, identifying the semantic foreground of the image becomes the first problem to be addressed In a deep CNN, the first convolutional layers learn low-level features, lines, dots, curves etc. while the other layers learn high-level features, objects and larger shapes in the target image. Since the semantic foreground must be analyzed against the high-level information of the image, we need to extract the high-level information of the image for processing. In the second step, the result of the first step needs to be precisely estimated, and we also need to extract the low-level information of the middle image and improve the fine-grained alpha map. For the above analysis, our architecture must be capable of doing the following: extracting highlevel semantic information and preserving lowlevel details, colors, textures, etc. That is the reason A-U-Net [14] is our preferred primary architecture. This framework covers the highlevel information of the image in the bottleneck and also covers the low-level information of the image in the shallow part of the encoder. In the bottleneck of A-U-Net, we add Atrous Spatial Pyramid Pooling (ASPP) [12] and a channel-wise attention-block [11] to extract the semantic foreground of the image. Through this design, the model initially accomplishes the first step mentioned above For the second step, we add skip connections between the encoder and decoder layers. We find that positional information is useful for both low-level structural information and high-level semantic information, so we add a spatial attention block to both low and high layers.

At this stage we can get an alpha matte with a good fine-structure border, but the alpha matte is prone to errors internally when the foreground has a cluttered texture or the foreground and the background have similar textures To alleviate this problem, we design an additional classifier to learn the pure foreground from the input image. Then the problem is solved successfully by combining the classification result and the A-U-Net result The details of our architecture are as follows

3.1 AU-Net

Base network. As shown in the blue box of Fig 2, our base network consists of four modules: encoder residual blocks [19], ASPP [12], and decoder. There are five convolution layers in the encoder and decoder each Each layer implements a 3×3 convolution, batch normalization [28] and ReLU [29] activation Bilinear operation is used for downsampling in the encoder and upsampling in the decoder. The decoder network skips connections with each layer of the encoder. There are four residual blocks [19] with dilated convolutions in the bottleneck Following the residual blocks, we adopt an ASPP module as used in the DeepLabV3 [12] approach. The ASPP module consists of multiple dilated convolution filters with dilation rates of 3, 6 and 9.

Attention block CBAM [11] is adopted in the architecture. As mentioned before, it utilizes both average and max pooling. To merge these two results in channel attention, it uses a similar operation to that used in SENet [15] separately and adds them elementwise while in spatial attention, it concatenates average and max pooling results and fuses them by a 7×7 convolution operation as shown in Fig. 3(a). To increase the receptive field as well as keep local features intact, we divide the spatial-wise attention block into three convolution layers of kernel size 11, 7, and 3. This modification is very helpful in the localization of the structure. As observed in the blue box of Fig. 2, we obtain spatial-wise attention maps, denoted as "S" in the first and second layers to rescale the encoder feature maps.

3.2 Classifier

With only the A-U-Net architecture, the alpha matte is prone to making errors when the foreground has cluttered textures or when both the foreground and

background have similar textures. To alleviate this problem, we design an additional classifier as indicated in the yellow box of Fig 2. We choose ResNet-34 [20] and its pretrained parameters for our classifier backbone. To have a better discriminative feature representation, a second-order channel attention (SOCA) module [13] is added next to the backbone; furthermore, ASPP is added between the backbone and SOCA to increase the receptive field. The output of the classifier is a 0–1 mask where 1s are those positions

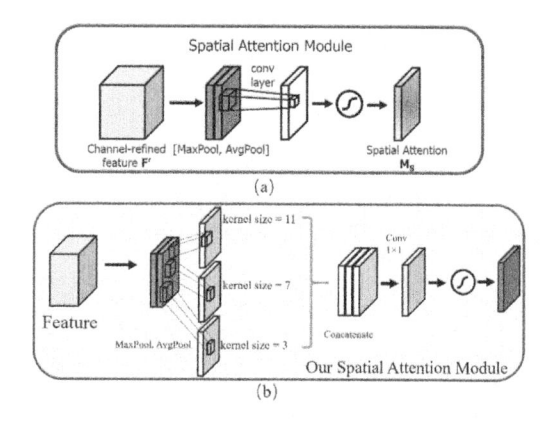

(a)

(b)

Figure 3. The proposed model. The whole architecture is divided into two parts: A-U-Net and a classification architecture with resnet-34 as the backbone. We added a residual block, ASPP, and a modified and specially designed CBAM block (including spatial-wise attention and channel-wise attention) to the A-U-Net. In the classification architecture, we added ASPP and SOCA, which can adaptively rescale the channel-wise features. H/W: the size of the feature; C: the channel of the feature.

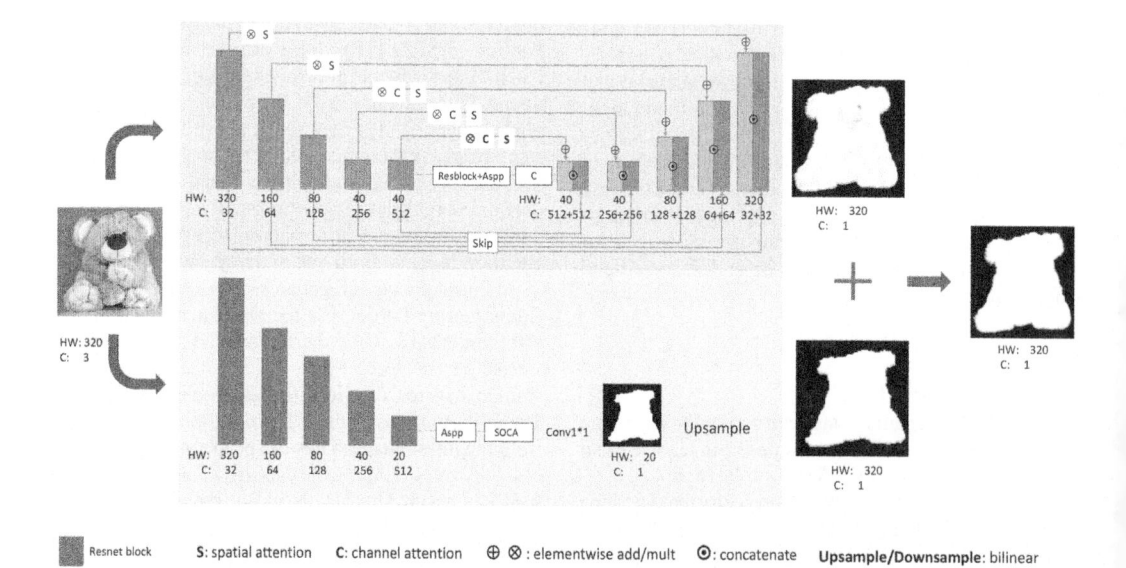

Figure 2. Spatial attention module. (a) is the original spatial attention module in the CBAM (b) is our modified module Instead of using one convolution operation of kernel size 7, we use three convolutions in parallel of kernel size 11, 7, and 3. The spatial attention map is obtained by implementations of concatenation, 1 x 1 convolution, and sigmoid function.

with alpha value 1 in the GT. Th intention is to identify the "pure white" area of the alpha matte.

3.3 Loss

The classifier module is designed keeping in view the following: 1-pure foreground, 0-background, and 2-opacity region. We use cross-entropy loss to train the classifier as

$$\mathcal{L}_{CE} = -\frac{1}{N} \sum_{i} \sum_{c=1}^{M} y_{ic} \log(P_{ic}), \tag{2}$$

where $M = 3$, y_{ic} is an indicator function which equals to 1 if the class of pixel i is c, and P_{ic} is the predicted probability of i belonging to c

Then, we generate a pure foreground mask X of size $k \times k$ based on the classifier results, i.e., the pure foreground parts are 1, and the remaining parts are 0:

$$I_{mask}(x_i) = \begin{cases} 1 & \text{if } \hat{y}_i = 1, \\ 0 & \text{otherwise,} \end{cases} \quad x_i \in X^{k \times k}, \tag{3}$$

where x_i is the value of pixel i of X and $\hat{y}_i$ is the predicted class of i

In the A-U-Net module, it is a typical regression problem to estimate the alpha matte and mean squared error (MSE) loss is commonly used. It can produce qualified alpha mattes by pixel-level supervision.

$$\mathcal{L}_{\alpha} = \frac{1}{N} \sum_{i} \left(\alpha_P^i - \alpha_G^i\right)^2, \alpha_P^i, \alpha_G^i \in [0, 1] \tag{4}$$

where α_P is the predicted alpha matte and α_G is the ground-truth alpha matte.

However, the alpha MSE loss only measures the difference in pixel space without considering overall structures of the foreground and background. Therefore, we introduce SSIM loss and total variation (TV) loss to calculate the structural similarity of the predicted foreground $\hat{F}$ to groundtruth foreground F.

$$\mathcal{L}_{ssim} = 1 - SSIM\left(\hat{F}, F\right) \tag{5}$$

where the predicted foreground is defined as $\hat{F} = I_{input} \times \alpha_P$ F is the ground-truth foreground.

Total variance of an image I is calculated as

$$TV(I) = \sum_{i,j} \left(\left(x_{i,j-1} - x_{i,j}\right)^2 + \left(x_{i+1,j} - x_{i,j}\right)^2\right)^{1/2}, \tag{6}$$

where $x_{i,j}$ is the pixel value on (i,j) position of I. In digital image processing, it is common to use TV loss

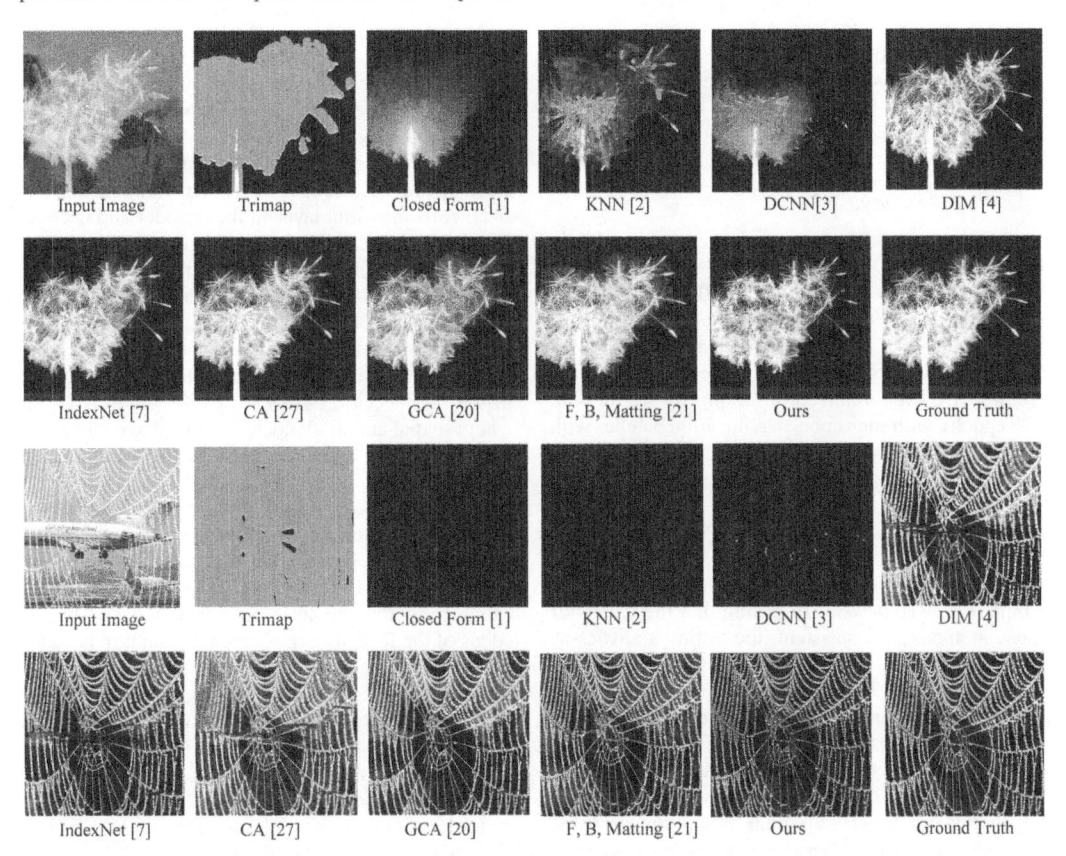

Figure 4. Qualitative comparison of the alpha matte results on the Adobe Composition-1k test set [4]. Our experimental results are in agreement with the results of [21]; finally the experimental results are cropped to a size of 800 × 800.

for denoising. In here, we use it to enhance structural similarity of predicted and groundtruth foregrounds as

$$\mathcal{L}_{fTV} = \| TV\left(\hat{F}\right) - TV\left(F\right) \|, \tag{7}$$

The complete loss function for training A-U-Net is given as

$$\mathcal{L}_{unet} = \lambda_1 \mathcal{L}_\alpha + \lambda_2 \mathcal{L}_{ssim} + \lambda_3 \mathcal{L}_{fTV}, \tag{8}$$

where λ_1, λ_2 and λ_3 are balance coefficients for loss

4 EXPERIMENTAL EVALUATION

We train our network using the public Adobe Deep Image Matting Dataset [4]. Various data augmentation techniques are used to avoid overfitting. We apply bionic transform, horizontal flip, blur, sharpen and random noise as data increments to the foreground, background and composition images. We compare our architecture with state-oftheart methods both quantitatively and qualitatively on the Composition-1k test dataset [4]. Four common metrics are used for evaluating the alpha mattes: sum of absolute differences (SAD), MSE, gradient error (Grad), and connectivity (Conn). The lower the above four metrics, the better the performance of the model. In addition, we conduct ablation studies on the Composition-1k test dataset to prove the effectiveness of different key parts.

4.1 Implementation details

For training, all composition images are randomly cropped to 320×320 640×640 and 800×800. Then they are resized to a resolution of 320×320. For loss optimization, we use the Adam. The learning rate is initialized to 0.01. The classification module is trained first. After convergence, we train the A-U-Net architecture. A-U-Net architecture training is set up with 100 epochs, with each epoch having 5000 batches with a batch size of 8.

4.2 Comparison to state-of-the-art

Composition-1k test dataset. We compared our method to different state-of-the-art methods on the Composition-1k dataset. To ensure that the results of those methods are consistent, the testing statistics as well as the image results are taken from the experimental results of [9]. Quantitative results are shown in Table 1. In comparison, our results are comparable to or even better than those of other state-of-the-art methods. Our three metrics SAD, Grad, and Conn are the best compared to other methods. Among those methods that do not use an additional trimap, the last three methods, our method performs the best in all metrics. Some image results for this dataset are shown in Fig 3.

Table 1. Comparison of the alpha matte with other methods on the Composition-1k test set [4].

Methods	SAD↓	MSE↓	Grad↓	Conn↓
Shared Matting [22]	125.37	0.029	144.28	123.53
Learning Based [23]	95.04	0.018	76.63	98.92
Global Matting [24]	156.88	0.042	112.28	155.08
Closed Form [1]	124.68	0.025	115.31	106.06
KNN Matting [2]	126.24	0.025	131.05	131.05
DCNN [3]	115.82	0.023	107.36	111.23
Information-Flow [25]	70.36	0.013	42.79	70.66
DIM [4]	48.87	0.008	31.04	50.36
AlphaGAN [5]	90.94	0.018	93.92	95.29
SampleNet [26]	48.03	0.008	35.19	56.55
Context-Aware [27]	38.73	**0.004**	26.13	35.89
IndexNet [7]	44.52	0.005	29.88	42.37
Late Fusion [18] †	58.34	0.011	41.63	59.74
HAttMatting [9] †	44.01	0.007	29.26	46.41
Ours†	**34.49**	0.012	**24.91**	**34.47**

† denotes that the method does not require a trimap as an additional input.

4.3 Ablation study

The ablation study for the architecture is carried out from the following two aspects: architecture and loss.

Architecture. Table 2 summarizes the contributions of each part of our framework to the overall framework. C and S denote the channel and spatial attention blocks respectively as suggested in CBAM [11]. S is our modified spatial attention block. "basic" denotes a basic A-UNet architecture consisting of an encoder and a symmetric decoder of five convolutional layers. Each corresponding layer in the encoder and decoder is skip connected. "basic+S" denotes an S is added in the shallow part (first two layers) of the encoder. "basic+S+<C_S>" denotes the channel- spatial-wise attention blocks added in the middle part (layers 3, 4, and 5) of the encoder. "basic+S+<C_S>+C" denotes a channel-wise attention block after ASPP is added in the bottleneck. "basic+S+<C_S>+C" is A-U-Net where spatial attention block S is replaced by the modified S. We use "A-U" to denote such an architecture in short.

In terms of A-UNet the performance is improved when spatial attention is added in the first two layers which helps localize the fine-structure on border edges. Additional <C_S> added in the middle layers helps grasp the semantic representation and border edges of the foreground. A channel attention is added after ASPP in the bottleneck further enhances the objectiveness of the foreground. Finally, except MSE, performances are largely improved when a modified spatial attention block is used.

In terms of the classifier structure, we compare the difference between using firstorder (C) and secondorder (SOCA) channel attention blocks which is to be added after ResNet. From Table 1, A-U-Net with a classifier can significantly enhance the performance. "Classifier_SOCA" is even better.

Table 2. Ablation study on the Composition-1k test set [4]. Those missing values mean the resulted alpha mattes are of poor quality.

Architecture	Alpha			
	SAD↓	MSE↓	Grad↓	Conn↓
Basic	–	–	–	–
basic + S	–	–	–	–
basic + S + <C_S>	55.43	.046	35.12	55.78
basic + S + <C_S> + C	53.29	.035	32.87	52.87
basic + S + <C_S> + C	48.44	.038	28.36	48.51
A-U+ Classifier_C	38.43	.016	31.35	38.20
A-U+ Classifier_SOCA	34.49	0.012	24.91	34.47

Various architectures are explained in the text.

Loss. In the loss ablation experiment, only losses in A-U-Net will be studied since the classifier is trained separately. Recall that three losses, MSE, SSIM, TV, are used in training A-U-Net. From the results illustrated in Table 3, we can see that SSIM loss has the greatest impact on the performance.

Table 3. The effect of each loss function in training the A-U-Net.

A-U-Net Loss	Alpha			
	SAD↓	MSE↓	Grad↓	Conn↓
MSE	49.70	0.16	35.13	47.17
MSE + SSIM	39.34	0.13	27.54	35.61
MSE + SSIM + TV	**34.49**	**0.012**	**24.91**	**34.47**

4.4 Limitations

The estimation of alpha values on a transparent foreground has been very challenging. As shown in Fig. 4, there are transparent cups superimposed on a background image. Although the classification result is correct, the resulted alpha matte is overprocessed. It gives an enhanced version of the GT.

Our estimated alpha matte heavily relies on the performance of the classification framework. Even though we are now able to achieve 97.3% accuracy for the classification framework, there is still a chance for error predictions. If the classification framework gives an incorrect prediction, this error will be magnified in the resulted alpha matte since the predicted "pure white" is to be upsampled to the original size before adding to the A-U-Net result. Fig. 5 illustrates the undesired noises in the alpha matte due to the misclassification of an opacity area as a pure foreground area. To improve the robustness of the framework, we may add the classified results to the middle layers of A-U-Net to lessen the scaling effect.

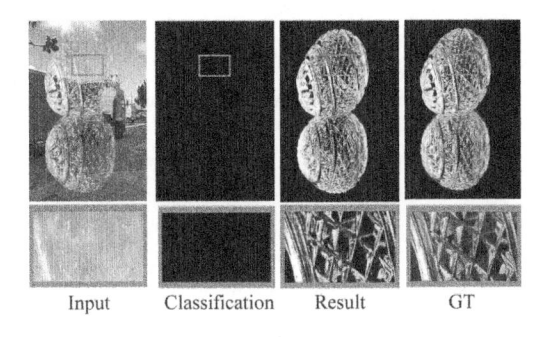

| Input | Classification | Result | GT |

Figure 5. The overprocessed alpha matte. Our result is more sharpened than that of the GT.

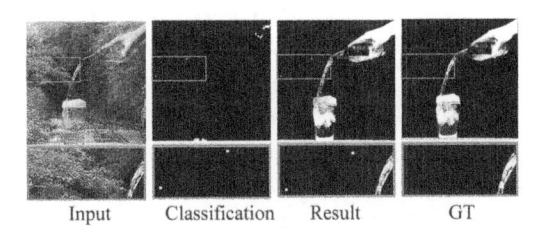

| Input | Classification | Result | GT |

Figure 6. Incorrect classification produces an alpha matte with noises.

5 CONCLUSION AND FUTURE WORK

In this paper, we design an architecture for image matting that can predict high-quality alpha matte from a single RGB image without additional inputs. It contains a main A-U-Net structure with the first-order channel and spatial attention modules and a classification structure with the second-order attention module. Our method is comparable to or even better than other state-of-the-art methods when viewed on a public test dataset.

The addition of the classification module resolves the problem related to the foreground and background having similar structures. Although utilizing classification can greatly improve the performance, it may also bring in limitations sometimes. Redundant "pure white" dots may appear when the classifier makes erroneous predictions. This will be our future work.

REFERENCES

[1] Levin, Anat, Dani Lischinski, and Yair Weiss. "A closed-form solution to natural image matting." IEEE transactions on pattern analysis and machine intelligence 30.2 (2007): 228–242.

[2] Chen, Qifeng, Dingzeyu Li, and Chi-Keung Tang. "KNN matting." IEEE transactions on pattern analysis and machine intelligence 35.9 (2013): 2175–2188.

[3] Cho, Donghyeon, Yu-Wing Tai, and Inso Kweon. "Natural image matting using deep convolutional neural networks." European Conference on Computer Vision. Springer, Cham, 2016.

[4] Xu, Ning, et al. "Deep image matting." Proceedings of the IEEE conference on computer vision and pattern recognition. 2017.

[5] Lutz, Sebastian, Konstantinos Amplianitis, and Aljosa Smolic. "Alphagan: Generative adversarial networks for natural image matting." arXiv preprint arXiv:1807.10088 (2018).

[6] Cai, Shaofan, et al. "Disentangled image matting." Proceedings of the IEEE/CVF International Conference on Computer Vision. 2019.

[7] Lu, Hao, et al. "Indices matter: Learning to index for deep image matting." Proceedings of the IEEE/CVF International Conference on Computer Vision. 2019.

[8] Chen, Quan, et al. "Semantic human matting." Proceedings of the 26th ACM international conference on Multimedia. 2018.

[9] Qiao, Yu, et al. "Attention-guided hierarchical structure aggregation for image matting." Proceedings of the IEEE/CVF Conference on Computer Vision and Pattern Recognition. 2020.

[10] Li, Jizhizi, et al. "End-to-end Animal Image Matting." arXiv preprint arXiv:2010.16188 (2020).

[11] Woo, Sanghyun, et al. "Cbam: Convolutional block attention module." Proceedings of the European conference on computer vision (ECCV). 2018.

[12] Chen, Liang-Chieh, et al. "Rethinking atrous convolution for semantic image segmentation." arXiv preprint arXiv:1706.05587 (2017).

[13] Dai, Tao, et al. "Second-order attention network for single image super-resolution." Proceedings of the IEEE/CVF Conference on Computer Vision and Pattern Recognition. 2019.

[14] Ronneberger, Olaf, Philipp Fischer, and Thomas Brox. "U-net: Convolutional networks for biomedical image segmentation." International Conference on Medical image computing and computer-assisted intervention. Springer, Cham, 2015.

[15] Hu, Jie, Li Shen, and Gang Sun. "Squeeze-and-excitation networks." Proceedings of the IEEE conference on computer vision and pattern recognition. 2018.

[16] Fu, Jianlong, Heliang Zheng, and Tao Mei. "Look closer to see better: Recurrent attention convolutional neural network for fine-grained image recognition." Proceedings of the IEEE conference on computer vision and pattern recognition. 2017.

[17] Mnih, Volodymyr, et al. "Recurrent models of visual attention." arXiv preprint arXiv:1406.6247 (2014).

[18] Zhang, Yunke, et al. "A late fusion cnn for digital matting." Proceedings of the IEEE/CVF Conference on Computer Vision and Pattern Recognition. 2019.

[19] He, Kaiming, et al. "Deep residual learning for image recognition." Proceedings of the IEEE conference on computer vision and pattern recognition. 2016.

[20] Li, Yaoyi, and Hongtao Lu. "Natural image matting via guided contextual attention." Proceedings of the AAAI Conference on Artificial Intelligence. Vol. 34. No. 07. 2020.

[21] Forte, Marco, and François Pitié. "$ F $, $ B $, Alpha Matting." arXiv preprint arXiv:2003.07711 (2020).

[22] Gastal, Eduardo SL, and Manuel M. Oliveira. "Shared sampling for real']time alpha matting." Computer Graphics Forum. Vol. 29. No. 2. Oxford, UK: Blackwell Publishing Ltd, 2010.

[23] Zheng, Yuanjie, and Chandra Kambhamettu. "Learning based digital matting." 2009 IEEE 12th international conference on computer vision. IEEE, 2009.

[24] He, Kaiming, et al. "A global sampling method for alpha matting." CVPR 2011. IEEE, 2011.

[25] Aksoy, Yagiz, Tunc Ozan Aydin, and Marc Pollefeys. "Designing effective inter-pixel information flow for natural image matting." Proceedings of the IEEE Conference on Computer Vision and Pattern Recognition. 2017.

[26] Tang, Jingwei, et al. "Learning-based sampling for natural image matting." Proceedings of the IEEE/CVF Conference on Computer Vision and Pattern Recognition. 2019.

[27] Hou, Qiqi, and Feng Liu. "Context-aware image matting for simultaneous foreground and alpha estimation." Proceedings of the IEEE/CVF International Conference on Computer Vision. 2019.

[28] S. Ioffe and Christian Szegedy. Batch normalization: Accelerating deep network training by reducing internal covariate shift. ArXiv, abs/1502.03167, 2015.

[29] V. Nair and Geoffrey E. Hinton. Rectified linear units improve restricted boltzmann machines. In ICML, 2010. 4

System Innovation in a Post-Pandemic World – Kin-Tak Lam et al. (Eds)
© 2022 Copyright the Author(s), ISBN: 978-1-032-24392-4

Research on the sound effects in the PILI puppet show—taking Tan Wu Yu of the PILI Bing Feng Jue in the drama martial arts as an example

L.-Y. Kuo

Smart Machinery and Intelligent Manufacturing Research Center, National Formosa University, Huwei, Yunlin, Taiwan
Department of Multimedia Design & Institute of Digital Content and Creative Industries, National Formosa University, Huwei, Yunlin, Taiwan, ROC

S.-Y. Guan & S.-T. Shen

Department of Multimedia Design & Institute of Digital Content and Creative Industries National Formosa University, Huwei, Yunlin, Taiwan, ROC

ABSTRACT: The purpose of this research is to explore the sound effects of the PILI puppet show and summarize the characteristics of martial arts.

In this paper, we take PILI Bing Feng Jue: Battle Torches-Match Point Chapter 25, Scene 3 Tan, Wu-Yu's exit martial arts as an example, conduct a case study of the sound effects of this scene, and analyze the sound effects. Then conduct expert interviews with experts in the company's sound team, such as mixing and sound effects, and verify the conclusions of both parties to find out the characteristics and rules of the relationship between the image and the sound, and the sound effects.

According to research results, it is found that sound of a cloth is used the most in terms of sound effect tone and usage count, followed by swoosh weapon, and simple delay is ranked third in usage count. The result shows that the sound effect is mainly cloth sound, combined with the body bump, which is the fourth most used, to express "Kung Fu." Swoosh weapon, matched with metal attack, also favors the sound of metal texture. The content of this martial arts uses a variety of "sword" cold weapons. Simple delay is used for the effect sound produced by key character actions or object movements that need to be emphasized in the plot to express "Qigong." The focus of this martial arts show is Tai Chi Xuan. The three elements of "Qigong," "Sword," and "Kung Fu" are the characteristics of the sound effects of the play.

It is hoped that through this research, practitioners will be able to more accurately grasp the key points of PILI puppet show sound effects, save costs, and have broader development in the future.

Keywords: Martial arts, Puppet show, Sound effect

1 INTRODUCTION

Since 2007, I have been engaged in sound effects (FX) work in the PILI Puppet Show (PILI) sound team, and before that, I had been exposed to PILI as a fan for many years. From these years of experience, I have found that there are far more researches on images than on sound.

In fact, the importance of sound in PILI is no less than the importance of the plot, images, and special effects. Among them, the research on the sound part is related to the uttered words and background music research, and the research on the FX part is rarely mentioned.

The purpose of this research is to study the FX of PILI, analyze the composition of the FX, and conduct expert interviews with experts who are also the staff of the PILI International Multimedia Production Team on mixing and FX. By verifying the conclusions of both parties, we hope to find out the images and sounds, and the characteristics and laws of the association.

2 LITERATURE REVIEW

PILI content is initially divided into literary and martial arts; compared to martial arts, FX in literary drama are mostly used to embellish the atmosphere. We will, however, not discuss it here. This article focuses on the FX of PILI martial arts.

This section first defines martial arts, and then discusses FX.

2.1 Definition of puppet show martial arts

According to Mr Jiang Wu Changs (1995) article "A Brief History of Taiwan Puppet Show," the development of Taiwan Puppet Show is divided into eight periods:

1. Cage bottom play
2. Beiguan Opera
3. Ancient books play
4. Swordsman play
5. Japanization
6. Anti-Communist and Anti-Russian
7. Golden Light Puppet Show
8. Radio and TV Puppet Show

From the period of swordsman play (1920–1936) in Taiwan puppetry, puppetry began to pay attention to the performance of various peculiar sword moves and the display of martial arts.

During the Golden Light Puppet Show period (1950, after the implementation of martial law), Puppet Show added a large number of colorful changing bulbs, artillery, smoke powder, smoke, stage lighting special effects, and various agencies to perform the sound and light special effects of swords.

Lin Shuyuan (2012) mentioned how Zhong Ren Xiang transformed real Kung Fu into puppet movements to increase the reality of puppet martial arts.

Radio and television puppet shows (from 1961 to present) are broadcast on radio and television. FX also became part of the puppet show during this period.

Cai Guorong (1983: 70) roughly divided martial arts action movies into three categories: "God and Monsters," "Swords," and "Kung Fu." The first two categories are collectively called "Martial Arts Movies."

According to Xu Jingwen's (2004: 17), classification of martial arts forms defines martial arts scenes as ones in which the characters in the play show Qigong, fist and Kungfu, or use swords and utensils in a play with more than two persons. Within this range, she divided the forms of martial arts into external and internal forms. Internal forms, such as Qigong or martial arts secrets, are used to increase internal power; external forms can be divided into three categories: fists, swords, and utensils.

2.2 *Definition of sound effects*

Sound can be divided into three parts: uttered words, background music, and FX in the postproduction process of PILI.

A dictionary definition of "sound effect" reads: "A sound other than speech or music made artificially for use in a play, film, or other broadcast production" (New Oxford Dictionary 1998).

Flückiger (2009 : G151) mentioned:

"In order to analyze and discuss FX he has developed a very simple modular framework for the description of FX, based on five questions regarding the source of the sound as well as its acoustic shape.

What is Sounding?

What is Moving? (The process-related nature of sound origination)

What Material is Sounding?

How Does it Sound?

Where Does it Sound?"

Flückiger (2009:153) also stated, "Sound is essentially related to movement. Movement only originates sounds."

Materials originating sounds are especially easy to describe.

According to Feng Xi (2015), movie sound is usually divided into three levels: meaning level (language, dialogue, and commentary), emotional level (passive music, ideographic FX, and environmental FX), and event level (active music and active FX).

In the event layer, the type of FX that is most closely integrated with the dynamics of the image can be called "vector sound effect (vector FX)," which has rich changes and diverse categories.

A film plays an important role in plot fluctuations, emotional rendering, space construction, sensory experience, and other aspects.

In the book *Practical Media Aesthetics—Image, Sound, and Motion* by Herbert Zetter, visual vectors are classified into three main types: graphic vectors, indicator vectors, and motion vectors. The FX accompanying the motion vector is the vector FX.

Feng Xi (2015) divides vector FX commonly used in movies into 11 categories and several subcategories:

1. Displacement

    This kind of vector FX corresponds to the effect sound produced by the motion trajectory of a single object in space over time, and is widely used. According to the specific situation, it can be subdivided into air breaking, friction, collision, crossing (Doppler), and other special displacement effects.

2. Ray

    This kind of vector FX corresponds to the effect sound produced by the high-speed linear motion of light or similar light.

3. Rotating

    This type of vector FX corresponds to the FX produced by the rotation of machinery or other objects. The action of rotation can be looped and is sustainable, so this type of FX is not transient but consists of loops of audio elements.

4. Zoom

    This type of vector FX corresponds to the effect sound produced by the expansion or contraction of an object.

5. Blasting

    This type of vector FX corresponds to the sound of explosions, destruction, or shock waves.

6. Cracks

    This kind of vector FX corresponds to the effect sound produced by irregular cracking of the object itself or by external force.

7. Dispersion

This kind of vector FX corresponds to the effect sound produced by a large number of (often the same kind) objects scattered on part or all of the screen and moving. This kind of dispersion is a visual special effect, and the corresponding FX must match the width and density of the spatial sound image and expand the visual depth tension.

8. Vibration

This type of vector FX corresponds to the effect sound produced by the rapid vibration of the object up and down or left and right.

9. Flashing

This kind of vector FX corresponds to the effect sound produced by the flashing of light on the object or the flashing of the button lights of the machine instrument and the graphic characters on the screen.

10. Liquid

This type of vector FX corresponds to the sound of liquid dripping, or the sound of an object falling into liquid. Therefore, it can be subdivided into two types: dripping and entering water, according to common situations.

11. Delay

This type of vector FX uses a delay effect device to make the effect sound of a vector repeated many a times. It is often accompanied by a slow motion lens. It is used for the effect sound generated by the key character action or object movement that needs to be emphasized in the plot. It can also be called "echo."

3 RESULTS AND DISCUSSION

3.1 *Research questions*

This study starts from the perspective of FX, discusses the characteristics, and summarizes the rules of the FX of the PILI. The research questions are the following two points:

1. Is it possible to describe the FX through text narration to find out the characteristics of the FX of PILI?
2. Is it possible to summarize the connection between the video and FX of the PILI through expert interviews and the above two points?

3.2 *Research methods*

This study uses case study to discuss individual cases. This research uses PILI Bing Feng Jue: Battle Torches-Match Point (BFJ) Chapter 25, Scene 3 Tan, Wu-Yu (Tan) exiting martial arts as a case for discussion. The time code on the DVD is 11:46 to 16:42.

The content of this scene is three-on-one with Jie Ao Zhu (Jie), Ye Fei Tian (Ye), and Cui Luo Han (Cui). Miao Feng Yun (Miao) strategizes in FengYun Liu Mian (Fen). The contrast between one movement and one silence, one danger and calmness is the rhythm of this martial arts show.

This scene is for Tan to exit the martial arts, and it is also an important stage for Tan to show real skills.

In terms of case selection, because of the above-mentioned reasons, this martial arts show is of great significance in the PILI.

3.3 *Research steps*

3.3.1 *Case study*

Because the researcher is involved in the FX of PILI, we can study directly with the sound track of the FX instead of separating the sound track.

This article divides BFJ Chapter 25, field 3 FX into seven segments according to the continuity of sound.

In this study, the FX are classified according to Feng Xi (2015) and refer to Flückiger (2009¡G151) to describe the timbre and sound through text narratives and arrange them (Table 1).

Table 1. The FX are classified according to Feng Xi (2015), and refer to Flückiger (2009 : G151) to describe the timbre and sound through text narratives, and arrange them.

No.	FX	Timbre
1	Cloth sound	Noise caused by cloth rubbing.
2	Body bump	The sound of physical impact
3	Swoosh weapon	The sound of weapons swinging, the material is metal, and the audio frequency is higher.
4	Metal attack	The sound of metal percussion.
5	Shoes run	The sound caused by friction between shoes and the ground.
6	Wind lite	The sound of the wind, the environmental sound of the wilderness.
7	Shoot sound	Qigong or sword qi emission, the effect sound produced by high-speed linear motion.
8	Spin sound	The effect sound produced by Qigong or self-rotating movement. The action of rotation can be looped and is sustainable, so this type of sound effect is not transient, but consists of loops of audio elements.
9	Pass through	The effect sound produced by light or Qigong expansion or contraction.
10	Magic spell	A large number of (often the same kind) objects are scattered on part or all of the screen, and the effect sound produced by the movement. This kind of dispersion is a visual special effect, and the corresponding sound effect must match the width and density of the spatial sound image and expand the visual depth tension.
11	Explosion boom	The effect of explosion, destruction or shock waves.

(continued)

Table 1. Continued.

No.	FX	Timbre
12	Spall granule	The object itself or the effect of external force cracks irregularly, resulting in sound effects.
13	Lightning thunder	The effect sound produced by the flashing or flickering of light.
14	Liquid pour	Liquid dripping, or the sound of an object falling into the liquid.
15	Simple delay	Use a delay effect device to repeat the effect sound of a vector many a times, often with a slow motion lens; used for the effect sound produced by the key character action or object movement that needs to be emphasized in the plot. It can also be called "echo."

In this study, the FX of this field are classified according to Table 1, and the number of times the FX are used is organized in Table 2.

Table 2. FX usage count.

No.	FX	Usage count
1	Cloth sound	76
2	Body bump	30
3	Swoosh weapon	66
4	Metal attack	12
5	Shoes run	19
6	Wind lite	6
7	Shoot sound	8
8	Spin sound	16
9	Pass through	20
10	Magic spell	23
11	Explosion boom	25
12	Spall granule	12
13	Lightning thunder	13
14	Liquid pour	4
15	Simple delay	37

It can be seen from Table 2 that cloth sound is used the most, followed by swoosh weapon, and the number of times of simple delay is ranked third.

The above results show that the FX of PILI is mainly cloth sound, which includes the sound of waving sleeves, punching, and fluttering cloth. This study judges that the material of PILI is cloth, and it is used with the fourth number of times. The body bump, the sound of the impact of the body represents "Kung Fu."

The second is swoosh weapon, which contains the sound of various weapons swinging, and with the metal attack, the sound is also biased toward the metal texture. This study judges that a variety of "swords" cold weapons are used in the content of the thunderbolt puppet show.

The third most frequently used is simple delay, which is used for the effect sound produced by key character actions or object movements that need to be

emphasized in the plot. This study judges that the focus of this scene is (Tai Chi Xuan; Martial arts moves).

Spin sound, pass through, and magic spell are all expressions of Qigong. This study judges that the PILI martial arts is one of the characteristics of using "Qigong." Explosion boom contains a variety of explosions. This study judges that PILI martial arts are mostly used to show the power of moves and exchanges.

From here, it is concluded that "the PILI martial arts includes the above-mentioned three elements of "Qigong," "Sword," and "Kung Fu."

3.3.2 Expert interview

The interviewees of this study are mainly members of the PILI sound group. The basic information of the interviewees is detailed in Table 3.

Table 3 Basic information of interviewed experts.

Table 3.

Expert code	Gender	Title	Seniority	Introduction
A	Male	Sound Mixer	15	He studied Recording Engineer at Fukuoka School of Music & Dance in Japan, and has about 5 years of experience in the Band.
B	Male	Sound Mixer	13	Previously engaged in recording equipment business.
C	Female	Sound Designer	20	At the same time working in PILI and as a Guzheng teacher
D	Male	Sound Designer	19	A fan of PILI, has no previous experience.
E	Male	Sound Designer	11	Served as an editor, and worked as a 3D stereoscopic design and correction for the movie (The Arti: the Adventure Begins), and then entered the sound group. Have experience in mixing and recording for a Band.

In the design of the interview topic, the two main frameworks are "background information" and "basic information about the FX of PILI." The opinions of experts are collected, and the basic information about the FX of PILI and martial arts is sorted and summarized. The research results in the previous section are verified. Expert interview topics are shown in Table 4.

3.3.3 Interview results and collation

From the interviews with the interviewed experts, it is known that the interviewed experts are all relevant workers who have been in the industry for many years, which is enough to strengthen the credibility of

Table 4. Topics of expert interview.

Background information
1	Where did you graduate from? Do you have any special experiences?
2	How long have you been in the related industry of sound group?
3	What is the reason for you to choose this industry?
4	What are your initial thoughts on the FX of PILI?

Basic information about the FX of PILI and martial arts
1	What do you think is the importance of FX of PILI martial arts?
2	What do you think are the main elements in PILI martial arts?
3	What kind of FX do you think are most commonly used in PILI martial arts?
4	Do you agree that "Qigong," "Sword," and "Kung Fu" are the main elements of PILI martial arts?

this interview. Among them, the interviewed experts also mentioned that the main elements in the PILI are "Qigong," "Sword," and "Kung Fu." The research results in the previous section have been verified.

In addition to the above three points, the interviewed experts also mentioned that in the past, the content of martial arts was more common in "Kung Fu" and "Sword." In recent years, because of the continuous development of computer animation, the proportion of "Qigong" martial arts has also increased. The puppets are mostly made of fabric, so the sound of cloth plays a huge part in puppet shows. Because the puppets are not performed by real people, some actions will be more abstract. At this time, it is necessary to rely on FX to make this action real. Simply put, it is to use FX to make up for the lack of picture and increase the credibility of the picture.

4 CONCLUSION

According to the research results, it is found that cloth sound is used the most in terms of FX tone and usage count, followed by swoosh weapon, and simple delay is ranked third in usage count. The result shows that the FX is mainly cloth sound, combined with the body bump, which is the fourth most used, to express "Kung Fu." Swoosh weapon, matched with metal attack, also favors the sound of metal texture. The content of these martial arts uses a variety of "Sword" cold weapons.

Simple delay is used for the effect sound produced by key character actions or object movements that need to be emphasized in the plot to express "Qigong." The focus of this martial arts show is Tai Chi Xuan. The three elements of "Qigong," "Sword," and "Kung Fu" are the characteristics of the FX of the play.

From the above research we are able to answer the following two research questions raised before.

1. We can describe the sound effects through text narratives to find out the characteristics of the sound effects of PILI puppet show.
   The main elements in the PILI puppet show martial arts are "Qigong," "Sword," and "Kung Fu."
2. Through expert interviews and the above two points, we can conclude the connection between the image and sound of the PILI puppet show.
   The conclusion of Question 1 is verified through expert interviews, and it is reliable.

This article only uses one case for research. It is inevitable that there are insufficient samples. In the future, more cases can be studied and more general conclusions have been obtained.

REFERENCES

Cai Guorong, 1983, "Yesterday, Today and Tomorrow of Dream Heroes—The Evolution and Influence of Martial Arts Movies", Literature and Art Monthly, No. 163, pp. 68–77.

Feng Xi, 2015, "On the Classification and Design Thinking of Vector Sound Effects", Modern Film Technology.

Flückiger, Barbara (2009). Sound effects: strategies for sound effects in film. In: Harper, G; Doughty, R; Eisentraut, J. Sound and Music in Film and Visual Media: an Overview. New York, USA: Continuum, 151–179.

Jiang Wuchang, 1995, "A Brief History of Taiwan Puppet Show", National Taiwan Art Education Center, Taipei.

S. Judy Pearsall, Patrick Hanks, 1998, "The New Oxford Dictionary of English", Oxford University Press.

Lin Shuyuan, 2012, "Research on the Inheritance of Puppet Play Skills in Xinxing Pavilion of Zhong Renbi", Master's thesis in Chinese Literature, Department of Chinese Language and Literature, National Kaohsiung Normal University.

Si Wu Yao, 2021, "JUST AMAZING talks about the peak of no desire, Tai Chi Xuan reappears!", Pili Monthly Magazine, Fuck a puppet 94 praise! , April 2021 April VOL.308, pages 32-34.

Xu Jingwen, 2004, "The Transformation of Chinese Martial Arts Movies under Globalization: Taking "Crouching Tiger, Hidden Dragon" and "Hero" as Examples", Master's thesis in Mass Communication, Tamkang University.

System Innovation in a Post-Pandemic World – Kin-Tak Lam et al. (Eds)
© 2022 Copyright the Author(s), ISBN: 978-1-032-24392-4

Improving the cooling rate of metal injection molding by implementing conformal cooling channels and a baffled hole system

Wei-Chen Zhou, Pei-Pu Song & Jiing-Yih La*
Department of Mechanical Engineering, National Central University, Taoyuan, Taiwan

Yao-Chen Tsai, Ming-Hsuan Wang & Chia-Hsiang Hsu
CoreTech System (Moldex3D) Co., Ltd., Hsinchu, Taiwan

ABSTRACT: Metal injection molding (MIM) is a kind of metalworking in which fine powder metal is mixed with binder material to create a feedstock that is then shaped and solidified using injection molding. As the metal powder and binder of an injected part are weakly bound during the cooling stage, the part temperate must be cooled down appropriately before the mold can be opened. It may be difficult for conventional two-dimensional (2D) cooling channels to deal with the problem of nonuniform temperature distribution for multicavity molds. In this study, we integrate the technologies of mold flow analysis, metal additive manufacturing and CNC machining for the design and manufacturing of a multicavity mold for MIM injection parts using a combination of conformal cooling channels and a baffled hole system. Mold flow analysis was implemented for the design of a conformal cooling and baffled hole system. Metal additive manufacturing and CNC machining were combined for the manufacturing of a core and cavity that has 3D cooling channels embedded inside. A four-cavity mold was manufactured to produce an MIM part, and a shop floor testing of this mold was performed to verify the quality of the injected part. In this study, we successfully demonstrate how to improve the cooling rate of a thick MIM part with an appropriate design of the cooling system. Also, a description of the integrated mold design and manufacturing technology is provided.

Keywords: Metal injection molding, conformal cooling channel, metal additive manufacturing, mold flow analysis, Feature recognition, inner-faces recognition, extrusion recognition, thin-wall parts

1 INTRODUCTION

Metal injection molding (MIM) is a kind of metalworking in which fine powder metal is mixed with binder material to create a feedstock that is then shaped and solidified using injection molding [1]. Unlike plastic injection molding in which the molecules of the injected part are strongly bound, powder metal and binder (such as wax) of the injected part (typically called green part) are weakly bound [2]. A process called sintering must be carried out on the green parts to evaporate wax and to bind metal particles into a coherent unit [3]. As the powder metal and binder of a green part is weakly bound during the cooling stage, the process parameters must be set appropriately; otherwise, substantial defects such as shrinkage, crack, and breaking may develop on the injected part.

Conventional mold design often employs two-dimensional (2D) cooling channels owing to their ease in machining. However, for a multicavity mold with a thick part on each cavity, 2D cooling channels may not be appropriate as the cooling channels can only surround the outside of all cavities, and cannot cover the high-temperature area of each cavity. Thus, it may be difficult to bring down the temperature appropriately during the cooling stage, which results in the formation of defects on the injected part. In contrast, when conformal cooling channels are employed, they can lie closer to the high-temperature area of the cavity, and hence the area can be cooled down more easily [4, 5]. However, conformal cooling channels cannot be fabricated by conventional CNC machining as they are curved shape and embedded inside a block material. The recent development of metal additive manufacturing (MAM) has made the fabrication of conformal cooling channels possible [6]. However, the entire core and cavity of a mold must be fabricated using the MAM technology. As the surface quality of an MAM part achieved through using MAM technology is still not satisfactory, the feasibility of this technology for application in precision mold manufacturing must be investigated.

The purpose of this study is to design and manufacture a four-cavity injection mold with conformal cooling channels and a baffled hole system to produce

---

*Corresponding Author

DOI 10.1201/9781003278474-7

a thick MIM part. The MIM part was originally fabricated by a one-cavity injection mold with 2D cooling channels. The main objectives for the development of the new mold are as follows: (1) increasing the number of cavities from one to four, (2) reducing the maximum temperature at the end of the cooling stage, and (3) reducing the cycle time for the entire process. The proposed mold design and manufacturing method integrates mold flow analysis, MAM fabrication, and CNC machining. Mold flow analysis is employed for the design of conformal cooling channels and a baffled hole system, whereas MAM fabrication and CNC machining are combined for the manufacturing of the core and cavity. The injected parts of the proposed mold are compared with those of a one-cavity mold with 2D cooling channels. The advantages of the proposed mold are discussed.

2 DESIGN AND MANUFACTURING OF THE MIM MOLD

Figure 1(a) depicts the CAD model of an MIM part investigated in this study, which has a dimension $28.8 \times 26.5 \times 28.8$ mm. The original mold is a one-cavity mold heated up and cooled down by 2D cooling channels. Figure 1(b) depicts the core and cavity design, and Figure 1(c) shows 2D cooling channels. The diameter of the cooling channels is 8 mm. There is only one layer of parallel channels across the neighborhood of the core and cavity. Figure 1(d) depicts the assembly of the core, cavity, and cooling channels, where one pair of inlet and output ports is designed for the core and cavity each. The problem of this molding process is that the cooling rate near the inner cylindrical surface is too slow for the cooling channels being located far away from it. If the temperature of the inner cylindrical surface is too high during the ejection, sticking on the cavity or core may occur.

Figure 2 depicts the flowchart of the proposed mold design and manufacturing process. In the mold design process, first, the CAD models of the core and cavity are designed through mirroring the original core and cavity for a one-cavity mold, respectively. Then, the distribution of the four cavities and the gates and runners must be redesigned. As conformal cooling channels are essentially 3D curves, they must be drawn segment by segment. Reference [7] provides guidelines for determining the position of the cooling channels, such as the minimum distance between neighboring channels, and the minimum distance between a channel and the part. Parts of the mold other than the core and cavity are also designed. Complete CAD models of the mold are assembled and evaluated. Engineering drawings of the core, cavity, and all mold plates are also provided, with tolerances specified on each part for the use in machining.

Next, in the mold flow analysis, the parameters to be set are divided into three types: machine parameters, material parameters, and process parameters. Most of the parameters are the same as those of an existing 2D mold used to produce the proposed MIM part, which can ensure that the simulation results do not deviate from the real situation too much. During the mold flow analysis, all results that affect the quality of the MIM part are analyzed, such as volume shrinkage, shear stress, shear rate, powder concentration, black line, total weight, and time required in different stages of the process. In particular, several types of cooling-channel design are investigated to analyze the temperature distribution and cooling time required, because the primary task is to improve the cooling rate of the high-temperature area of MIM. The types of cooling channels studied include 2D cooling channels

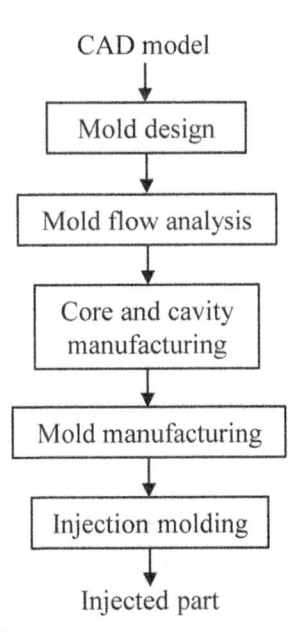

CAD model

↓

Mold design

↓

Mold flow analysis

↓

Core and cavity manufacturing

↓

Mold manufacturing

↓

Injection molding

↓

Injected part

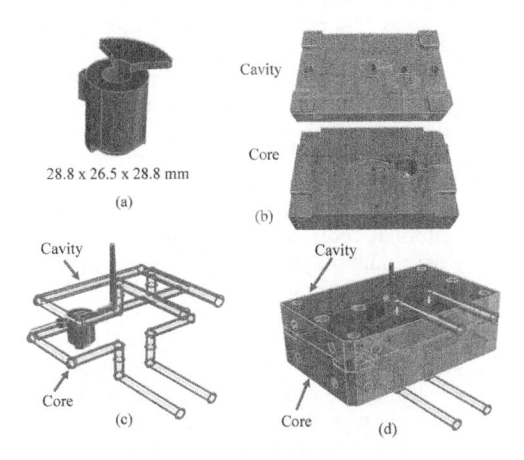

Figure 1. Original one-cavity mold design for the MIM part. (a) CAD model, (b) cavity and core design, (c) 2D cooling channels, and (d) assembly of the core and cavity.

Figure 2. Flowchart of the proposed mold design and manufacturing process for MIM parts.

(one-cavity mold, 8 mm in diameter) and conformal cooling channels (two types of 3D contour design, 4 and 6 mm in diameter, respectively). The diameter of 2D cooling channels is larger than that of conformal cooling channels because the interference may occur for the latter if its diameter is too large. Finally, conformal cooling channels with a diameter of 4 mm are selected for the final design.

The next step is core and cavity manufacturing. The core and cavity are manufactured by OPM250L—an MAM machine. Each of the CAD models is converted into an STL model first. The STL model is offset outside to leave a thickness of 0.5 mm for the secondary machining with CNC machines. The offset STL model is then used by OPM250L to produce an AM part. Two AM parts are fabricated for the core and cavity each. The metal powder used in OPM250L is 420J2. The AM part must be heat-treated by tempering to eliminate residual stress. The surface hardness after tempering is about 51 ± 1 HRC. The surface quality of the AM part is actually very rough. The secondary machining includes CNC milling, electrical discharge, wire electric discharge, and polishing. After the secondary machining, the surface accuracy and roughness can reach 0.01 mm and 0.5 μm Ra, respectively, thus fulfilling the requirements of the MIM molding application.

In mold manufacturing, the mold base, including all kinds of mold plates, components, and parts, is manufactured. The entire mold, including the core, cavity, and mold base, is then assembled and modified. Finally, the mold is tested on the shop floor with an Arburg injection machine. The metal powder used in the test is Fe_2Ni. The mold is tested and repaired three times to eliminate all defects found.

3 RESULTS AND DISCUSSION

The CAD model of the injected part is shown in Figure 3(a), which essentially includes four parts: one runner and three flash grooves. The runner and flash grooves are loosely connected to the parts so that they can easily be removed after ejection. Figure 3(b) depicts the assembly of the mold designed in this study and its dimension, where the core, cavity, all mold plates, and components are assembled. Figure 3(c) and (d) depict the CAD models of the core and cavity, respectively. The thicknesses of the core and cavity are 35 and 20 mm, respectively. It is to be noted that 3D cooling channels are embedded inside each of these two components. Figure 3(e) depicts the cross-sectional view of 3D cooling channels on the core and cavity, which have three layers and one layer, respectively. The inlet and outlet ports of the cooling channels on each plate are also indicated. Figure 3(f) depicts the 3D view of the assembly of the core, cavity, ejectors, and 3D cooling channels. The most challenging part of 3D cooling channel design is that there is not enough space for the placement of cooling channels. If the distance between any of the neighboring channels or a channel

and the part is too small, the failure rate in MAM fabrication may be increased. Therefore, a trial-and-error approach is employed to design 3D cooling channels manually.

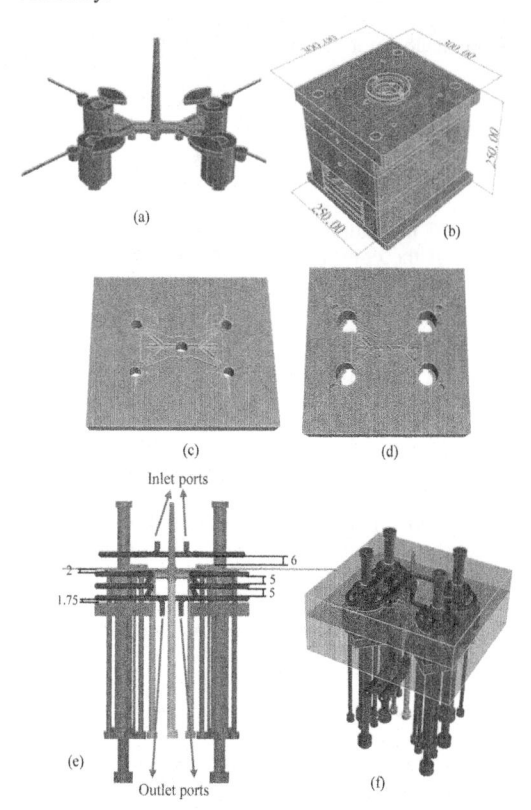

Figure 3. The proposed four-cavity mold for an MIM part. (a) The injected part, (b) the mold set, (c) the core, (d) the cavity, (e) conformal cooling channel design (front view), and (f) a 3D view.

For mold flow analysis, several types of cooling channels have been designed and tested. Figure 4 depicts three of them, which are 2D cooling channels (Figure 4(a)), conformal cooling channels (Figure 4(b)), and conformal cooling channels plus a baffled hole system (Figure 4(c)). Two-dimensional cooling channels are primarily used for the comparison with conformal cooling channels. Both the core and cavity have one layer of channels, with a diameter 8 mm. The total surface area of all cooling channels is 26831 mm^2. For conformal cooling channels, both the core and cavity have multiple layers of channels, each with a diameter 4 mm. The cooling channels surround each core and cavity as close as possible. The total surface area of all cooling channels is 30272 mm^2. A baffled hole system is added to the conformal cooling channels to enhance the cooling of four inner cylindrical surfaces on the parts. It is an individual cooling system and has its own inlet and outlet ports. The diameter of the baffled hole system is 4 mm. Moldex3D was used in mold flow analysis and the above-mentioned three types of cooling channels were simulated.

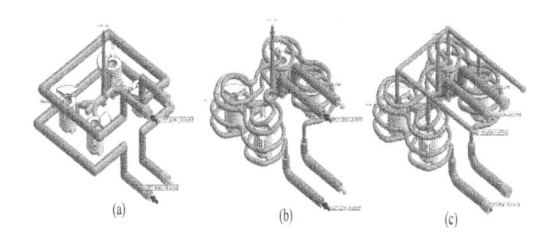

(a) (b) (c)

Figure 4. Three types of cooling channels tested in mold flow analysis: (a) 2D cooling channels, (b) conformal cooling channels, and (c) a combination of conformal cooling channels and a baffled hole system.

Figure 5 depicts the temperature distribution of the part at the end of the cooling for three types of cooling channels, where the cooling time is set to be 20 s. For 2D cooling channels (Figure 5(a)), the maximum and minimum temperatures are 87.2°C and 60.4°C, respectively; for conformal cooling channels (Figure 5(b)), the maximum and minimum temperatures are 85.3 and 59.6°C, respectively; for the combination of conformal cooling channels and a baffled hole system (Figure 5(c)), the maximum and minimum temperatures are 68.5°C and 60.1°C, respectively. This result shows that the temperature distributions for conformal and 2D cooling channels are quite similar, indicating that conformal cooling channels alone cannot produce satisfactory results. In contrast, when conformal cooling channels and a baffled hole system are combined, the maximum temperature can be brought down to 68.5°C from 87.2°C.

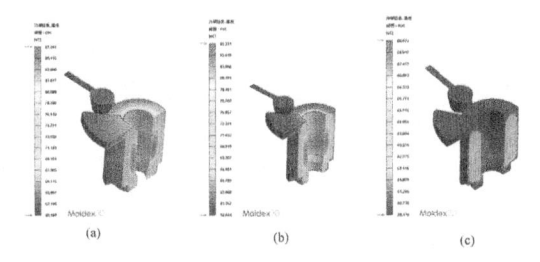

(a) (b) (c)

Figure 5. Temperature distribution for three types of cooling channels: (a) 2D cooling channels, (b) conformal cooling channels, and (c) a combination of conformal cooling channels and a baffled hole system.

Figure 6 depicts the time required for the three types of cooling channels to reach the temperature of ejection (96°C for this kind of material). The maximum time required for 2D cooling channels, conformal cooling channels, and a combination of conformal cooling channels and a baffled hole system are 14.7, 14.2, and 7.6 s, as shown in Figure 6(a), (b), and (c), respectively. Again, this result shows that the combination of conformal cooling channels and a baffled hole system is the best option among the three cooling systems. Therefore, this option is selected and implemented on the proposed mold.

Figure 7(a) and (b) show the MAM parts of the cavity and core, respectively. It can be seen that the surface quality is very rough, as expected. However, what we are really concern with is the surface quality of the part after fine machining. Figure 7(c) and (d) show the parts of Figure 7(a) and (b), respectively, after the secondary machining. After checking the surface quality of two parts in detail, we found no defect on the surface. Therefore, such MAM parts can be used in molding applications. A local region of the part surface in Figure 7(d) is expanded to illustrate the surface quality.

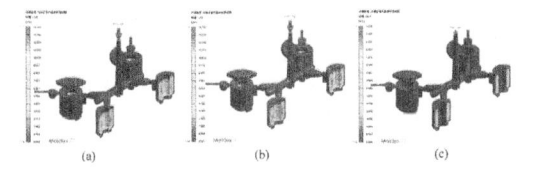

(a) (b) (c)

Figure 6. Time required to reach the ejection temperature for three types of cooling channels: (a) 2D cooling channels, (b) conformal cooling channels, and (c) a combination of conformal cooling channels and a baffled hole system.

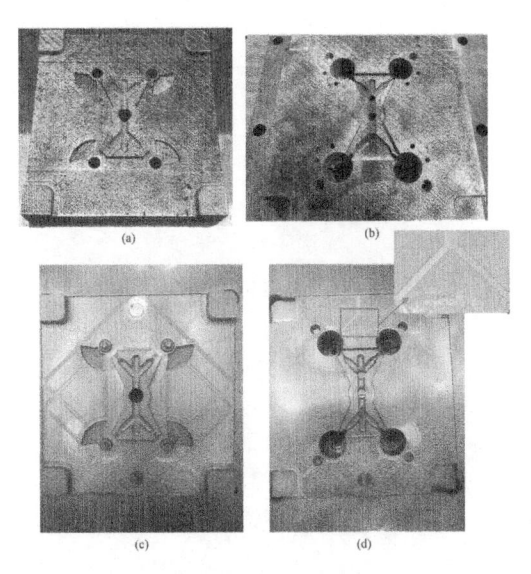

(a) (b)

(c) (d)

Figure 7. Manufacturing of the core and cavity: (a) MAM part of the cavity, (b) MAM part of the core, (c) the cavity after the secondary machining, and (d) the core after the secondary machining.

Figure 8(a) and (b) depict the male and female molds, respectively. All cooling channels are inside the molds, and hence not visible from the outside. It is to be noted that conformal cooling channels exist only on the core and cavity. Two-dimensional cooling channels must still be manufactured on the core and cavity plates, and the transition between a 2D channel and a conformal channel is sealed by an O ring. Several defects owing to inaccurate machining and alignment were observed during the injection test. However, after several rounds of repairing and adjustment, the quality of the final injected parts becomes satisfactory. During the injection test, the adjustment of the process parameters is an important factor. Two main defects were observed during the setting of parameters: shrinkage on inner cylindrical surface and breaking on the runner. However, these defects can be overcome by adjusting the molding pressure, molding time, feeding volume, and V/P. Final injected parts of the test are shown in Figure 9.

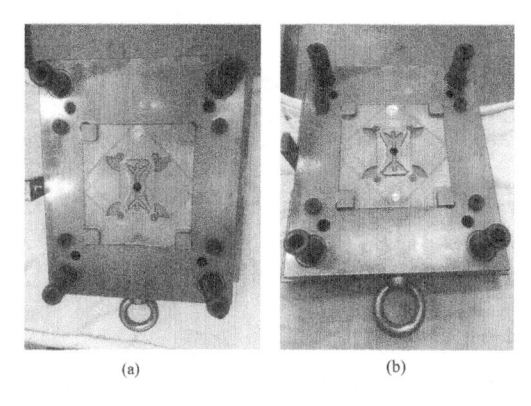

(a)   (b)

Figure 8. (a) Male and (b) female molds.

Figure 9. Final injected parts.

4 CONCLUSION

In this study, a four-cavity MIM injection mold was designed, manufactured, and tested. A combination of conformal cooling channels and a baffled hone system was developed to overcome the slow cooling rate of the inner cylindrical surface of an MIM part. Mold flow analysis was performed on three types of cooling channels, and the one with a combination of conformal cooling channels and a baffled hone system has been shown to be most efficient in terms of lowering the temperature at the end of cooling and the time required to reach the ejection temperature. MAM and conventional CNC machining processes were employed in producing the core and cavity. MAM was employed to fabricate the core and cavity that have conformal cooling channels embedded inside. CNC machining was then implemented to fine-tune the surface quality of the core and cavity. An entire mold set was manufactured in this study, and extensive injection tests were performed. Satisfactory injected parts were finally obtained after appropriate tuning of the process parameters during the mold test. The main contribution of this study is t the integration of conformal cooling channels design, mold flow analysis, MAM, and CNC machining for the design and manufacturing of a multicavity mold for an MIM injection part. We successfully integrated all technologies to produce an MIM mold qualified for real-life applications.

REFERENCES

[1] Ing. Frank Petzoldt, Powder Injection Moulding International 8, 2014, 37–45.
[2] Marco Thornagel, Metal Powder Report, 65, 2010, 26–29.
[3] S.C. Hu and K.S. Hwang, Powder Metallurgy, 43, 2000, 239–244.
[4] E Sachs, S Allen, M Cima, E Wylonis, and H Guo, Production of Injection Molding Tooling with Conformal Cooling Channels Using the Three Dimensional Printing Process, Polymer Engineering and Science, 40(5), 2000, 1232–1247.
[5] K.W. Dalgarno and T.D. Stewart, Manufacture of Production Injection Mould Tooling Incorporating Conformal Cooling Channels via Indirect Selective Laser Sintering, Proceeding of the Institution of Mechanical Engineers, Part B, 215, 2001, 1323–1332.
[6] "Laser Additive Manufacturing: Going Mainstream" https://www.osa-opn.org/home/articles/volume_28/february_2017/features/lasenstream/
[7] Whitepaper, Conformal Cooling Using DMLS, GPI Prototype & manufacturing Services.

System Innovation in a Post-Pandemic World – Kin-Tak Lam et al. (Eds)
© 2022 Copyright the Author(s), ISBN: 978-1-032-24392-4

Virtual reality: A new method for smart aviation maintenance training services

Van Hoan Vu* & Wen-Chung Wu
Department of Aeronautical Engineering, National Formosa University, Hu-Wei, Yunlin, Taiwan

Liang-Yin Kuo
Smart Machinery and Intelligent Manufacturing Research Center, National Formosa University, Hu-Wei, Yunlin, Taiwan

ABSTRACT: With the rapid development of technology in recent years, virtual reality is becoming increasingly popular in many different fields, especially in the education sector. This article presents a virtual system that uses virtual reality devices for pre-flight checks for the Dornier 228 aircraft, representing a smart aviation maintenance training method besides traditional training methods. In the near future, this system will be tested in training at the Aviation Maintenance Training Center, National Formosa University. All reviews, comments, and suggestions will be recorded for evaluation and comparison of this training method using virtual reality technologies with other existing state-of-the-art training methods. By using the virtual reality-powered training method, the motivation level and effectiveness of trainees could be significantly enhanced. The experiment shows that virtual reality has great potential not only in teaching and training but also in motivating students and enhancing their interaction with the lesson.

1 INTRODUCTION

Aviation is one of the five industries in the field of transportation. This is the most modern transportation industry with a high level of safety, convenience, and speed. Along with the advancements in science and technology, its development has been deemed necessary with ever-increasing human travel and freight needs. Thus, for the development of society and the economy, the technological advancement of this industry has become essential. Maintaining the transportation process to be safe and fast is a critical task. This is achieved through performing many processes, including inspection and maintenance, from simple to complex. However, the traditional maintenance method is gradually becoming obsolete and has many problems associated with it. In light of the recent advancements made in the field of science and technology, finding an advanced method to replace and improve the traditional training methods has become inevitable.

Virtual reality (VR) is an advanced technology that allows users to interact with 3D models in a virtual environment. Basically, VR has three important properties: immersion, interaction, and imagination (Zou Ying-zhi et al., 2010). Using them, a VR system can be evaluated and its applicability is determined. Some VR applications include design, manufacturing, entertainment, education, inspection, and maintenance.

Since its inception, there had been many studies conducted on the applicability of VR in different fields, but due to the lack of progress in science and technology, advancements could not be made at that time. VR also play an important role in the entertainment field, such as video games and 3D cinema. However, with the advancement of science and technology in recent years, outstanding benefits of VR could be gradually revealed along with its high applicability, especially in the fields of education and maintenance.

In the medical field, VR acts as a low-cost and easy-to-use training method for medical students (Aman S. Mathur, 2015; Mingwei Cai et al., 2019). In the military, soldiers can be trained in different situations and environments or in the maintenance of military equipment in a virtual environment using a VR device (Aaron Gluck et al., 2020; Pin Duan et al., 2012). In architecture, VR can prove to be a useful tool for the inspection and evaluation of a building's structure, thereby facilitating easy and effective maintenance (A. Zita Sampaio, 2012). Automotive is also one of the areas where VR technology can be used to its full potential, from design and production to inspection and maintenance (Song Wei, 2018). In aviation, there have been many studies conducted using VR for maintenance, training, and inspection (Xintong Shao et al., 2019; Xiaolin Quan et al., 2011) to develop a new training method for smart aviation maintenance besides traditional training methods.

*Corresponding Author

All this proves that VR will become the technology of the future with distinct advantages like lower costs and providing more safety than other traditional methods, which incur a high cost, especially in the aviation industry. Therefore, VR can prove to be the ultimate solution to satisfy the ever-increasing demands of modern aviation engineering education.

2 METHODOLOGY

In this paper, a virtual maintenance training system is designed with the aim of testing it in maintenance training at National Formosa University to evaluate the potential and feasibility of using VR as a new method for smart aviation maintenance training services.

For this system, the Dornier 228 aircraft is being chosen as a prototype for interaction in a virtual maintenance environment. To create a 3D model of the aircraft, the CATIA V5 software is used, which is widely used in the aviation industry (Figures 1 and 2).

Then, the Unity game engine is used as a core development environment to design a virtual environment. This game engine has many powerful tools to support building VR experiences. The C# language is used to write scripts that are used to trigger behaviors, and object and user interactions of the system. One of the most essential tools of this system is the VR device. A set of HTC Vive Pro is chosen for viewing and interacting in virtual environments. Vive Pro has an AMOLED display for richer colors and contrasts, enhancing the user experience.

Figure 1. Different 3D views of the Dornier 228 aircraft.

Figure 2. VR device kit of HTC Vive Pro.

3 EXPERIMENTAL SECTION

In this system, several scheduled maintenance "A Check" tasks are created such as inspecting the flaps, rudders, and propellers of the aircraft to interact with the 3D model in the virtual environment (Figure 3). Software Development Kit (SDK) is needed to enable VR devices to collect all input from the real environment and translate it into the virtual environment. Nowadays, there are many powerful and useful SDKs available for all types of devices, such as SteamVR Plugin from Valve Corporation or OpenVR Plugin from Unity Technologies, Oculus XR Plugin from Unity Technologies, and Oculus Integration from Oculus for Oculus's devices. Our system uses the SteamVR Plugin based on the excellent support of this SKD for Vive, and it is also the ultimate tool for experiencing VR content. The positions that perform tasks also offer teleport points to help users change positions easily. Besides instant teleportation methods, users can also move using the touchpad on the controller, making it more realistic.

Figure 3. The virtual training system of the Dornier 228-212 aircraft.

Some information about aircraft components, maintenance processes can be present as pictures for Figure 4. Hand tools and protective gears for maintenance training.

Figure 4. Hand tools and protective gears for maintenance training.

Moreover, the system can also use videos to make the virtual training program more effective.

A few tools such as Monkey Wrench, hand drilling, and pliers are added for training purposes. Moreover, safety is very important in aviation maintenance so safety helmets and ear defenders are added to enable students focus their attention on the training.

4 RESULT AND DISCUSSION

Through the experiment, we can see that this virtual training system can simulate the tasks of a real-life maintenance training program. Furthermore, all tasks are designed to be simple and easy to motivate students for the training, thereby enhancing their learning experience.

By applying VR in training and teaching, space and cost for the infrastructure and equipment could be significantly reduced. On the other hand, it can minimize the chance of occurrence of faults in the real equipment. It also enhances the student's learning experience and motivation, and they can learn about different parts of the equipment without actually seeing or disassembling the real equipment. Virtual training system can help overcome the disadvantages associated with traditional teaching methodologies (Xiaolin Quan et al., 2011). VR provides a virtual environment that enhances safety and reduces risk in the teaching process.

However, to exploit the full potential of VR, the requirements for an advanced hardware configuration to support it is a challenge facing us. Health concerns related to prolonged use of VR should also be kept in mind. VR device developers recommend users to avoid prolonged use of VR devices and take a break every 30 minutes to 1 hour (Figure 4).

5 CONCLUSION

VR is a technology with potential to replace traditional aviation teaching/maintenance practices with smart aviation maintenance training services to improve safety, engagement, and motivation for students.

In the future, we will test this system at NFU University to evaluate the feasibility of the system and collect data and student feedback to improve the system, and will also assess the impact of VR on human health with the aim of contribute to the existing literature on this topic for scientific purposes.

In light of the global impact of the COVID-19 pandemic, it is a challenge to demonstrate the potential of VR in distance learning and teaching.

REFERENCES

Aaron Gluck, Jessica Chen, Ratnadeep Paul, 2020. *AIVR*. 386–389.
Aman S. Mathur, 2015. *VR*. 345–346.
Mingwei Cai, Reika Sato, Kenji Yoshida, Kenta Takayasu, 2019. *NicoInt*. 5–7.
Pin Duan, Lei Pang, Yong Jin, Qi Guo, Zhi-Xin Jia, 2012. *ICQR2MSE*. 1396–1399.
Song Wei, 2018. *ICVRV*. 140–141.
William R. Sherman, Alan B. Craig, 2003. *Encyclopedia of Information Systems*. 589–617.
Xintong Shao, Xiaofei Wei, Shengxian Liu, 2019. *ICCASIT*. 147–150.
Xiaolin Quan, Phong, Limin Qiao, Shaochun Zhong, Shusen Shan, 2011. *MEC*. 5–7.
Zou Ying-zhi, Lv Chuan, Liu Rui, 2010. *ICALIP*. 1284–1288.
A. Zita Sampaio, Daniel P. Ros'rio, 2012. *CISIS*. 507–512.

System Innovation in a Post-Pandemic World – Kin-Tak Lam et al. (Eds)
© 2022 Copyright the Author(s), ISBN: 978-1-032-24392-4

Discussions on internet public voices and their influence on the popular animation film PUI PUI Molcar published over the internet

Chian-Fan Liou & Pei-Chi Su*
Department of Visual Communication Design, Southern Taiwan University of Science and Technology, Yongkang, Tainan, Taiwan

Shih-Chieh Liao
Department of Multimedia and Entertainment Science, Southern Taiwan University of Science and Technology, Yongkang, Tainan, Taiwan

Chao-Chih Huang
Department of Popular Music Industry, Southern Taiwan University of Science and Technology, Yongkang, Tainan, Taiwan

ABSTRACT: Digital Content Industry has brought a revolution to entertainment habits following the speedy development of internet hardware and software right after radio and TV media. The way people receive multimedia messages has changed from receiving messages-only to interactive messaging and sharing on the public platforms. As the internet information sharing platforms diversify, the enterprises and creators also need to change their strategies for marketing their creations and products as time changes. As most internet platforms have the character that records user's use patterns, it is easier to use mass data and internet public opinions to analyze the marketing performance of multimedia products, which is efficient and valuable referencing information. Given the fact that mass data sampling is nonjudgmental and uncontrollable compared to traditional manual surveys, this essay is going to observe and analyze how the popular animation film, PUI PUI Molcar, became a hit over internet by using the Semantic Differential Technique via OpView and following the animated film keywords, "PUI PUI" to gather and analyze information of public discussions of this animation over Facebook, YouTube, and Instagram. Additionally, these three major internet platforms in Taiwan can be used to observe how the multimedia products develop and impact people after being on the market as a reference for future enterprises and creators of multimedia products to learn the basic operation of internet marketing. Take "PUI PUI Molcar," for example, its first published video by far was mainly exposed on YouTube. However, only 85 voices were recorded in YouTube platform whereas 151 voices were heard in Facebook, which is regularly used in Taiwan and 287 voices were recorded in Instagram that is popular to young users. If we only assess the marketing performance based on public voices on one single platform, it is easy to misjudge and make a wrong decision for commercial operations. This essay is going to analyze and study this case in order to provide meaningful insights for enterprises, users, and creators who plan to publish their multimedia works or products on the internet platforms.

Keywords: Internet Public Voices, PUI PUI Molcar, Anime.

1 INTRODUCTION

With the rapid development of network software and hardware, Digital Content Industry has become one of the most promising industries in this century (S.-W. Yeh et al., 2007). In modern times, it has replaced the past radio and TV media, and has become a new habit for most people to obtain information and entertainment. With the characteristics of the media and the evolution of the market in recent years, the way people receive audio-visual information has changed from message-only to interactive behaviors such as commenting, messaging, and sharing. A survey conducted by the National Development Commission in 2017 showed that more than 89% of people in Taiwan use social software every day. They are social and audio-visual media lovers and are highly attached to those things (D.-Y. Zhou et al., 2019).

In Taiwan, the five most popular social media platforms involve 90% of overall internet users for public

---

*Corresponding Author

DOI 10.1201/9781003278474-9

voices, which are Facebook, YouTube, Instagram, and two forums, PTT and Dcard (OpView, 2020). Those social media platforms have become commercial advertising platforms for enterprises, markets, and individual users. Marketing strategies need to change as time moves. The amount of internet advertising in 2016 was greater than TV advertising. Soon enough, the internet advertising banners were replaced by social media advertising pop-up windows (Y.-S. Wu, 2017), which indicates the importance of internet social media's impact on market development.

Facebook accounts for more than 50% of the total internet public voices and hence becomes the main observation indicator. Although Instagram's public voice is not high, it has grown by 60% from 2019 to 2020 during the same period (OpView, 2020). YouTube has also become a self-advertising media platform. All three are important observation indicators for Internet users. As the epidemic affects countries and industries, the Internet economy has received strong attention. With the diversified competition of market products and brands, the massive information transmitting can easily lose consumers' interest and reduce their attention (OpView, 2021). To successfully promote products and brands, in addition to following the algorithmic mechanism of the Internet platform, studying consumers' usage habits and preferences of the target group and increasing adhesion can effectively achieve influence.

Scholars Mark Bonchek and Vivek Bapat also pointed out in the article: "The Most Successful Brands Focus on Users—Not Buyers." The brands marketed through the Internet in digital times are significantly different from traditional brands. People no longer understand brands through advertising and traditional messages. Instead, they build their feelings about the new digital brand through dialogs and sharing. The user's experience, evaluation, and feelings become indicators that affect future users' understanding of the brand and whether to use it. Therefore, the management of brands between digital times and traditional times is totally different. What one needs to consider is not potential customers, but future users (Mark Bonchek & Vivek Bapat, 2018).

2 ANALYSIS AND DISCUSSION

Before 2007, most of Taiwan's animation industry was still a technology company with a foundry nature, and only a few companies had early-stage planning capabilities. Compared with countries with more mature cultural and creative industries, they lacked the ability to commercialize creativity (C.-Y. Chen, 2011). For the future development strategy of the animation industry, if we refer to Japan, the world's largest exporter of animation and culture, the industry and company operating structure are relatively different from those in Europe and the United States. The type of industry in Japan is relatively similar to that of Taiwan. The "production committee" operates in a series mode, so

the research and analysis of popular online short animations released by individual small studios are more likely to be implemented, as it is low-cost and highly effective for most original companies at this stage. Scholars S.-S. Chen & H.-J. Lin have also put forward the same view, believing that the series of short films not only trains the overall industrial process to be smoother, but also arises the attention and response of the audience to achieve a more mature development (S.-S. Chen & H.-J. Lin, 2017).

In early 2021, a guinea pig-themed animated film went on the Internet in Taiwan, which became popular among social media, resulting in image copying of guinea pigs (such as car paint, handicrafts, etc.) and a trend of showing off guinea pigs (sharing pictures and videos of guinea pig pets) from all walks of life. There are even news and a large number of netizens worrying about the tragedy of guinea pig breeding and abandonment. "PUI PUI Molcar" is a children's program produced by the new Japanese director Misato Tomoki and released on TV Tokyo. Taiwan's edition was run by Muse Communication Co., Ltd, and the agency was released on the YouTube network platform. From January 5, 2021, after the release in Japan, Twitter followers reached more than 150,000 people in only 10 days, and it exceeded 407,000 on March 19. In Taiwan, beginning in March 2021, an official website was established to provide more information and authorization cooperation channels.

To produce short-length animations in multiple episodes to attract public attention and have them circulated through the internet, to implement brand licensing, and to create cross-industry alliances need very successful commercial operations. It has to break through the traditional animation media in TV or film Framework, but due to the chaotic and changeable network environment, even if it may become popular overnight, it may be quickly forgotten or replaced. Therefore, this article will use a large data of internet public voice surveys that cannot be manipulated to check the animation album's influence and trends in Taiwan after its breaking popularity and to provide a contextual reference for subsequent animation-related industries and creators' online distribution. This article will use OpView database to apply semantic analysis technology to track the internet public voice and analyze public opinions by searching the animation keyword "PUI PUI Molcar" over the three internet social media platforms, Facebook, YouTube, and Instagram. The analysis is as follows:

Between January 1, 2021, and April 1, 2021, the internet public voices published over the major three platforms were 287 records on Instagram, 151 records on Facebook, and 85 records on YouTube, as shown in Figure 1.

From Figure 1, we can see that this animated film only received 16.3% of internet public voices on YouTube although it was published originally on YouTube, which accounts for less than 20% of total public voices. However, Facebook received 28.9% public voices, nearly 30%, and Instagram

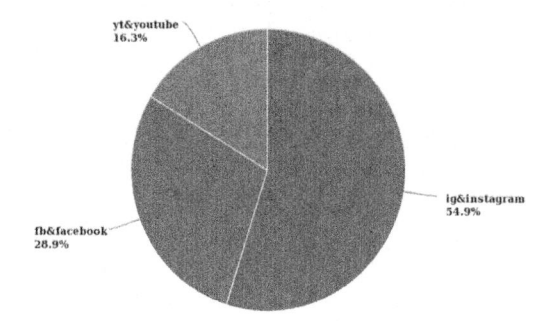

Figure 1. Distribution of internet voices on three major platforms.

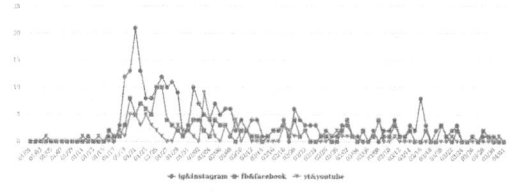

Figure 3. January 2021 to March 2021: The daily internet public voice change chart (horizon indicates the date and vertical means public voices).

received 54.9% public voices, which was over 50%. From this distribution, we learn that although this film was published on YouTube originally, this platform was not the main indicator to observe the popularity of internet public voices. We need to take into account Taiwan internet users' habits to discuss the observation indicator on film publication over the internet, especially on Instagram and Facebook. Particularly the cartoon style like PUI PUI Molcar, Instagram is the major market platform.

Internet public voices not only come from the three major social media platforms, but also from other intermedia, like forums, Q&A websites, news, social group websites, review websites, blogs, etc. over a diverse range. To verify that social media websites are the main indicators to observe the overall internet public voices, Figure 2 shows the analysis: Over 80% of internet public voices come from Instagram and Facebook, which are social media websites (the red bar) whereas 20% come from forums (yellow bar), news (blue bar), and blogs (green bar). Within YouTube, over 70% of internet public voices also came from social media websites (red bar). We should take notice that the News (blue bar) website accounts for a bigger percentage, 20% for public voice sources compared to the other two platforms, and its forum (yellow bar) also has a bigger percentage of internet pubic voices. However, there is no such big difference regarding the yellow bar and green bar compared to the other two platforms.

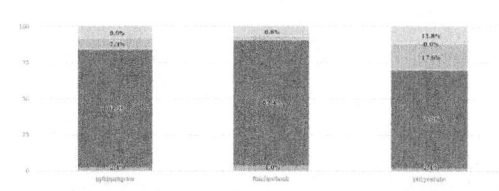

Figure 2. Percentage of public voice distribution (yellow bar: forums, blue bar: news, red bar: social media websites, green bar: blogs, purple bar: Q&A websites, and orange bar: review websites; the last two bars have nearly zero records.).

According to Figure 3, it is found that the overall public voices change over time. Based on official data of PUI PUI Molcar, Episode 1 was published on

January 5, 2021, followed by the next 11 episodes released every 7 days.

Table 1. "PUI PUI Molcar" release time

Episode	01	02	03	04	05	06	07	08	09	10	11	12
Date	1/05	1/12	1/19	1/26	2/02	2/09	2/16	2/23	3/02	3/09	3/16	3/23

From the chart, we can see that there was no internet public voice when Episode 1 was on. After Episode 3 was on, there was a sharp growth and then it dropped gradually. By observing the detailed peaks, we found out that the small peaks always took place a few days after the film was released and it dropped regardless of the published timing. The continuous peaks happened when Episodes 3 and 4 were on, which was at its best popularity. After Episode 5, the peak still remained quite high compared to the later episodes. However, the internet public voices only started to grow from Episodes 6 to 12 after they were launched. We can conclude the following points from the overall development:

1. The no-feedback time before the film was firstly launched was normal, as it took time to grow in popularity.
2. There was only a short window of time when the popularity started to spread, about after 3 to 5 episodes were released when the audience started to discover the film and talk about it.
3. The follow-up internet public voices only continued after the new episode was released.

From the average growth graph on the three major platforms, we can see that even if the public voices are not in the highest volume on YouTube, the film upload timing and initial broadcasting were almost immediate. On the other hand, it takes 1 to 2 days for Instagram and Facebook to catch people's attention with more powerful influence. From the graph of Instagram, we can see there was an unusual peak on 16 March, which was when the anime "Time machine PUI PUI Molcar" was on. This anime virtualized traveling through time in human history and guinea pig history to describe their evolution. In this anime, the real guinea pigs were also used as characters to play in some of the scenes. Whether it was these additional characters' (real guinea pigs) appearance or the story content of the anime leading to the higher volume of public voices encourages a deeper quantitative and qualitative investigation.

3 CONCLUSION

In terms of Taiwan internet public voices, the common information exchange platforms are as such: forums, news websites, social group websites, reviews websites, and blogs. In this essay, we discuss "PUI PUI Molcar" with a cute cartoon style as an example for its performance on the three major social media platforms: YouTube, Facebook, and Instagram. To observe the outbreak of popularity for this anime, Instagram is the indicator platform, followed by Facebook. YouTube is a different observation indicator platform for public voices compared to the other two. Although YouTube is a video upload platform, in Taiwan, most people still use Facebook and Instagram to exchange messages and information.

Based on the detailed public voice graph chart observation, we suggest that the author or film producer does not have to worry about the early comments on the first few episode launches as it is the attention-building period for people to notice the film and comment. But there is only a short period of time for increased public voices in a sharp growth, which only happens for 2 or 3 weeks. The producer should seize the timing to do more promotions on the films in big-scaled commercial operations to efficiently take advantage of the outbreak of increased public voices.

The keyword, "PUI PUI Molcar" (precise) was set for search in this essay for internet public voice through strict filtering due to the word limit setting. However, there are other short names for this anime, such as the commonly seen "PUI PUI" or "Molcar," which has not been included for observation analyses. The word used on the internet also affects the overall data performance. If there is a need to do further research in depth, it is sensible to consider the local language use habit on the internet for public voice observation to obtain more comprehensive data and more precise analyses.

REFERENCES

C.-Y. Chen, 2011, Cultural Limitations on the Development of the Taiwanese Animation Industry, Journal for Studies of Everyday Life., (4), 47–87.

S.-S. Chen & H.-J. Lin, 2017, Animation Industry Analysis and Research of Adaption from Picture Books to Animations and Investigation of Related Techniques, Journal of National Taichung University of Science and Technology, 4(1), 187–207.

Mark Bonchek and Vivek Bapat, 2018, The Most Successful Brands Focus on Users – Not Buyers, Harvard Business Review, 2018/02/07.

OpView Community Word of Mouth Database, 2020, Wanglu Yuching Jhengba Jhan Fensi Shih Gongkai Shangji Wajyueh Shu, 2020/10/29.

OpView Community Word of Mouth Database, 2021, Wanghong Singsiao Pinggu Jhihnan Dazao Pinpai Shechyun Chyuan Sin Jhanli, 2021/04/15.

Y.-S. Wu, 2017, Tachu Chengwei Wanghong De Diyibu Yonggan Ansia Jhihbo Jian Ba!, Zen Cosmos, 150, 38–45.

S.-W. Yeh, Y.-C. Syueh and D.-J. Hong, 2007, "Chian" Jin Shuwei Neirong Chanyeh Chuangzao Jingjheng Youshih – Jiehjing Rihben De Chanyeh Fujhih Moshih, Industry Management Review, 1(2), 17–28.

D.-Y. Zhou, S.-J. Kang, J.-Y. Lyu, H.-J. Sieh, 2019, Wanglu Hongren Kesin Du Yingsiang Siao Fei Jhe Taidu Jhih Yanjiou – Yi Youtube Biaoyan Lei Wanghong Weili, Tuwun Chuanbo Yishu Syuehbao, 87–101.

The impact of COVID-19 on the well-being of youth and the effectiveness of online learning in higher education—a Taiwanese perspective

Siu-Tsen Shen*
Department of Multi-media Design, National Formosa University, Hu-Wei, Yunlin, Taiwan

Stephen D. Prior
Aeronautics, Astronautics and Computational Engineering, The University of Southampton, Hampshire, UK

ABSTRACT: This study presents an overview of the COVID-19 pandemic and its impact on Taiwanese youth well-being, together with the effectiveness of online learning in higher education. Recent related research was reviewed which gave insight to the project goals. An online survey technique was deployed, based on empirical evidence, on the effects of the COVID-19 pandemic on youth mental health and the evaluation of their campus e-learning performance. The results showed that the participants did feel a certain degree of anxiety, depression, loneliness and reduced sleep quality. COVID-19 related worries about themselves, their friends and family was highly evident. However, the willingness to receive a vaccination was quite low (53.3%), probably due to the fact that Taiwan has to date had a low incidence of COVID-19 infection. Furthermore, the findings highlighted that the participants were satisfied with the online learning platform used during the lockdown and social distancing, but would prefer a blended learning environment post-pandemic. Future work will focus on PTSD-linked research and efforts towards mental health and psychological well-being of adolescents and young adults in a post-pandemic world, as well as the implementation of an effective cloud-based learning and teaching model.

Keywords: COVID-19; Youth mental health; Youth well-being; Higher education; Online learning.

1   INTRODUCTION

The COVID-19 (C-19) pandemic of 2020/21 has had a great impact on people's mental health. According to the World Health Organization's (WHO) report in May 2021, there have been 159,319,384 confirmed cases of C-19, including 3,311,780 deaths (WHO2021) to date (Organization 2021). Many more cases and the resulting deaths have sadly gone unreported, due mainly to health services in poorer countries being completely overwhelmed by this novel coronavirus. Governments around the world have taken strict measures to control the spread of the virus; measures that have restricted social contacts to the absolute minimum. The study presented in this paper has been designed to investigate the impact of the C-19 pandemic on adolescents and young people's mental health and psychological well-being in Taiwan.

This study aimed to determine the well-being and the effectiveness of online learning in higher education during the COVID-19 pandemic. This research was a quantitative descriptive study using survey methods conducted online. The sample of the study was from undergraduate and postgraduate students across universities in Taiwan. Primary data collection in this study was carried out by distributing questionnaire online to 407 respondents. Data obtained by filling out questions that were distributed to all respondents in Google Form via the link: https://forms.gle/AvjvoEj7LCeqBRsH6.

2   MATERIAL AND LITERATURE REVIEW

The United Nations Children's Fund (UNICEF) conducted a poll which showed that the C-19 crisis has had a significant effect on the mental health of adolescents and young people in Latin America and the Caribbean. As stated in the report, the rapid assessment amplified the voices of 8,444 adolescents and young people between the ages of 13 and 29 in nine countries and territories in the region. The report gave an account of the feelings they faced in the first months of the response to the pandemic and the situation in September 2020. One in two felt less motivated to do activities as they normally enjoy. Three out of four felt the need to ask for help regarding to their physical and mental well-being. Despite having felt the need to do so, two out of five did not ask for help (UNICEF 2021).

Among the participants, 27% reported feeling anxiety and 15% depression in the last seven days. For 30%, the main reason influencing their current emotions is

*Corresponding Author

DOI 10.1201/9781003278474-10

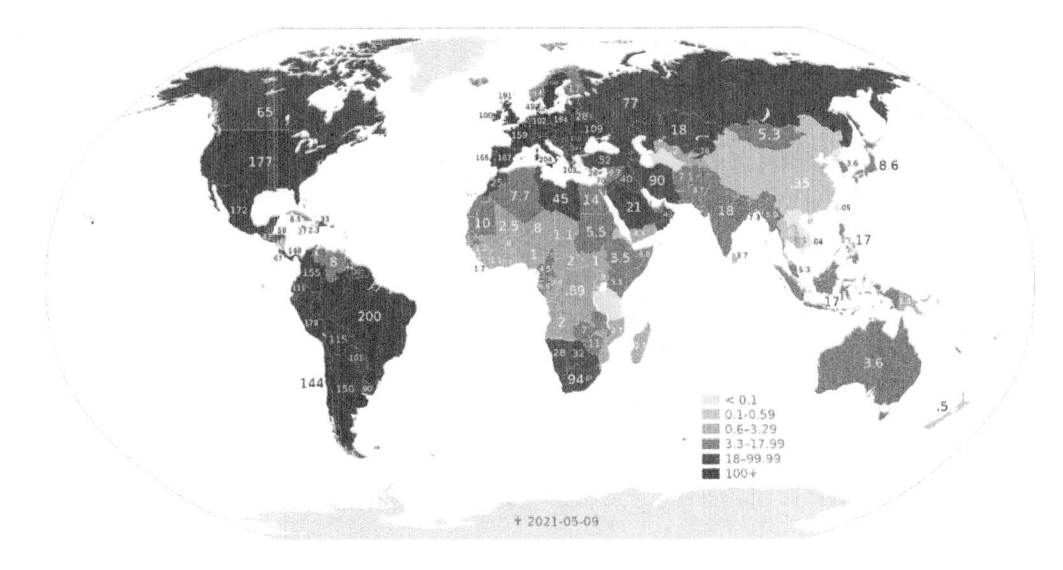

Figure 1. COVID-19 outbreak world map total deaths per capita (Wikipedia 2021).

the economic situation. A situation that generates deep concern and is a call to national health authorities is that 73% have felt the need to ask for help concerning their physical and mental well-being. Despite this, 40% did not ask for help (UNICEF 2021).

Huang and Zhao (2020) conducted a web-based cross-sectional survey, and collected data from 7,236 self-selected volunteers assessed with demographic information, C-19 related knowledge, generalized anxiety disorder (GAD), depressive symptoms, and sleep quality. The results showed that the C-19 outbreak did significantly affect the mental health of Chinese public, and young people had a higher risk of anxiety and depressive symptoms than older people. Furthermore, younger people spending too much time thinking about the outbreak, and healthcare workers were at high risk of mental illness (Huang and Zhao 2020).

Ahmed, Ahmed, Aibao, Hanbin, Siyu, and Ahmed (2020) investigated a sample of 1,074 Chinese people through an online survey, majority of whom from Hubei province, China. Results showed higher rate of anxiety, depression, hazardous and harmful alcohol use, and lower mental well-being than usual ratio. Results also revealed that young people aged 21–40 years are in more vulnerable position in terms of their mental health conditions and alcohol use (Ahmed, Ahmed et al. 2020).

Evans, Alkan, Bhangoo, Tenenbaum, Ng-Knight (2021) implemented longitudinal data to characterize effects on mental health and behavior in a UK student sample, measuring sleep quality and diurnal preference, depression and anxiety symptoms, well-being and loneliness, and alcohol use.

Self-report data was collected from 254 undergraduates (219 females) at a UK university at two-time points: autumn 2019 (baseline, pre-pandemic) and April/May 2020 (under 'lockdown' conditions). Longitudinal analyses showed a significant rise in depression symptoms and a reduction in wellbeing at lockdown. Over a third of the sample could be classed as clinically depressed at lockdown compared to 15% at baseline. Sleep quality was not affected across the sample as a whole. The increase in depression symptoms was highly correlated with worsened sleep quality. A reduction in alcohol use, and a significant shift towards an 'evening' diurnal preference, were also observed. Levels of worry surrounding contracting C-19 were high (Norbury and Evans 2019, Evans, Alkan et al. 2021, Evans and Norbury 2021).

Hansan and Bao (2020) used the Kessler psychological distress scale ($K10$) to evaluate stress symptoms. They conducted an online questionnaire amongst college students in Bangladesh. Results show that "e-Learning crack-up" perception has a significant positive impact on student's psychological distress, and fear of academic year loss is the crucial factor that is responsible for psychological distress during COVID-19 lockdown (Hasan and Bao 2020). Furthermore, Husky, Kovess-Masfety, and Swendsen (2020) discovered that two-thirds of university students in France experienced increased anxiety as well as moderate to severe stress during C-19 confinement. Respondents who did not relocate to live with parents were disproportionately affected (Husky, Kovess-Masfety et al. 2020). Chen, Liang, Peng, Li, Chen, Tang, and Zhao (2020) found that 7.7% of college students showed depressive symptoms during the lockdown. Unfavorable living rhythms were associated with depressive symptoms (Chen, Liang et al. 2020). Recent research by Janiri, Carfi and Kotzalidis (2021) has stated that Post-Traumatic Stress Disorder (PTSD) may occur in individuals who have experienced a traumatic event. Previous coronavirus epidemics were associated with

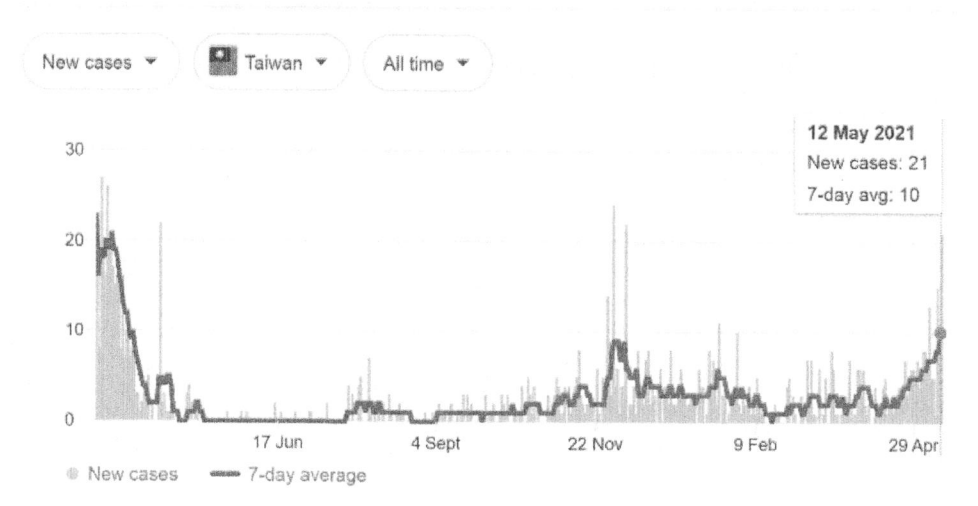

New cases and deaths

From <u>JHU CSSE COVID-19 Data</u> · Last updated: 1 day ago

Figure 2. COVID-19 case timeline in Taiwan (Data 2021).

PTSD diagnoses in post-illness stages, with meta-analytic findings indicating a prevalence of 32.2% (Janiri, Carfi et al. 2021).

3 RESULTS

The present survey was conducted online using Google documents through Line and Facebook with 407 participants over a period of two months (April-May 2021). The survey was conducted in three parts. Part I included demographics of the subjects, which consisted of seven questions that captured a general overview of the participants and their background (gender, age, ethnicity, working status, education, social media app use, etc.). Part II consisted of 16 questions related to the participant's health and well-being. Part III from students' perspectives concerning the factors influencing their intention to use e-learning system within the Taiwanese universities context, which consisted of eight main questions relating to their online experience of education, whilst being in lockdown during the pandemic.

3.1 *Part I – general overview review*

In terms of gender, the questionnaire participants were slightly more male dominated (57.2%) as opposed to female (43.5%). There was a broad range of ages represented, with the mean age being 22 years old. The survey being focused on Taiwan, the vast majority of the participants were Taiwanese (80%), with other nationalities being mainland Chinese (5.2%), Malaysian (4%), Japanese (2.7%), Pakistan (1.9%), Indonesian (1.7%), and a range of other nationalities with smaller percentages. Full-time students represented the major grouping (84%), followed by those

in full-time employment (12.3%). The remainder were made up of smaller percentages of self-employed, unemployed, etc.

Many of the participants were studying for a University degree (56%), Masters (28.7%), PhD (4.4%), of the remainder most were high school graduates. A fairly even mix were majoring in either Arts & Humanities (34.6%) or Engineering (33.7%), with a large range of other subjects, including Management (5.9%), Materials (4.7%), Business (3.7%), IT (2.9%), Education (2.5%), etc. The final question in this section asked about Social Media apps that they used most frequently. There was a wide range of answers, the most popular being Line, Facebook, YouTube, Instagram, Messenger, Google+ and Twitter. Of the others mentioned, WhatsApp, Pinterest and WeChat were also popular.

3.2 *Part II – personal emotional well-being*

When asked whether the participant's mental health had been adversely impacted by C-19, it was a surprise to find that on a five-point Likert scale, the highest response was (1) 'Not at all' (36.1%) with the rest on a sliding scale (2) 23.6%, (3) 20.4%, (4) 12.8%, with only 7.1% of the participants selecting (5) 'A great deal'. To some extent this is understandable given the relatively low impact to date of C-19 in Taiwan.

The following question asked about how well they adapted to isolation and social distancing. Given that Taiwan has not been under a prolonged lockdown and therefore few people have had to isolate for long periods, it was no surprise that the only 2.2% of the participants answered 'Not at all', the answers then increased (2) 6.1%, (3) 21.4%, (4) 33.7%, and then (5) 'Very well' 36.6%.

The following question asked how well do you feel you have been able to follow the government rules. Again, there was no surprise to find that most people (49.6%) were able to follow these with no problem, at the other end of the scale, only 0.5% found this very difficult. In terms of worries of developing C-19, (question 4) was answered Yes (67.8%), two thirds to one third. When asked about the worry of friends and family developing C-19 this increased to 80.6% 'Yes'. This clearly shows the level of fear of C-19 within Taiwan.

When asked whether the participant had ever had a C-19 test, only 7.4% said 'Yes'. Again, Taiwan has experienced a very low level of C-19 infection, with only 1,231 cases and 12 deaths (as of 12 May 2021) (CDC 2021). Of these 30 participants, only 2.4% (2) stated that the result was positive. Interestingly, only 18.2% isolated for the required 14 days. Taiwan recently began its vaccination program, so it was no surprise to find that only 1.2% of the participants had been vaccinated. When asked if those that had not been vaccinated would accept a vaccination, only 53.3% said that they would. Of those that wouldn't accept a vaccination if offered, most cited Trust Issues (127), Side Effects (116), Not Needed (58), Fear of Needles (22) and a small number citing a Religious Belief (participants could select more than one answer).

When asked if they knew of anyone that had contracted C-19, 89.2% said 'Yes'. Of this total 39.8% were admitted to hospital, and sadly 45.7% of those died.

When asked about their sleep quality, 57% had on average 8 hr/night. One quarter said that their sleep pattern was good, with 17% stating that they had poor sleep quality. Of those that had poor sleep quality, 50.7% had between 4–6 hr/night, 41.2% experiencing 6–8 hr/night. Rather surprisingly, 6.8% were surviving on 2–4 hr/night, and 1.4% on less than 2 hr/night.

When asked about what strategies people use to alleviate symptoms of poor sleep quality, there was a range of answers, ranging from Online Entertainment (244), Exercise (190), Talking (180), Trying to Stay Positive (177) and Volunteering (13). A very small number were actively seeking medical/professional counselling. (Note, participants could answer more than one category hence the total will not add up to 404).

When asked if they feel anxious, 60.4% said 'Sometimes' and 6.4% answered 'Often'. When asked about depression, 52.8% answered 'Sometimes', and 4.9% 'Often'. Clearly, there is a large percentage of the population with underlying fears and mental issues that they would not normally discuss.

When asked if people sometimes felt lonely, 59% stated that they sometimes did, with 10.1% stating that they often did. In terms of the participants' general health/fitness level, just over half stated that they thought they were about average, with 14% Poor and 2% Very Poor. When asked if they thought their general health/fitness had got worse in the last year, 13.5%

thought that it had, with 41% stating 'Maybe' (a level of uncertainty here, underlying how people perceive this, if maybe they don't regularly exercise).

When asked about whether they thought they were suffering from long COVID, a surprisingly high 56.8% stated that they were! This might be psychosomatic or 'lost in translation', given the low level of incidence of C-19 in Taiwan. However, a small percentage of participants were from overseas, and so they might have experienced an infection overseas and recovered before returning to Taiwan. A further explanation might be that they were asymptomatic and therefore might not have even known that they had had C-19.

When asked how long their long COVID had existed, there was a large range of answers, ranging between 7.1% and 17.9%; the highest percentage being for the 3-6 months' category (17.9%). Somewhat surprisingly, only 2.7% of the participants that thought they had long COVID stated that they were getting medical help with the condition. Therefore, it would appear that there is a large body of unreported and undiagnosed people suffering in silence, and dealing with this challenging condition on their own.

3.3 Part III – personal learning questions

When asked if their educational establishment had provided access to a remote online learning platform, 64.6% stated 'Yes'. One would have expected this to be higher. However, this again might be a factor of not having a long duration lockdown in Taiwan. In terms of which platform was provided, there was a range of answers, Zoom (93), Google Meet (74), Microsoft Teams (47), Discord (28), Vimeo (12), Skype (7) and Goto Meeting (2). There were a multitude listed in the Others category such as eCampus, etc.

In terms of satisfaction with the course progress, the participants, most were satisfied (56.2%). In terms of the e-classroom interaction, again there were high levels of satisfaction (48.5%). Likewise, in terms of the e-classroom ambience, we observed high satisfaction levels (48.3%). In terms of fairness, most participants were happy (60.3%). In terms of the teaching performance, again we observed high levels of satisfaction (62.7%).

In terms of the platform used, it would appear that overall the participants are satisfied with the level of functionality with (54.5%) in category 4 (Satisfied) and 5 (Very Satisfied). Likewise, in terms of stability of the platform, the participants rated this highly (52.1%). In terms of the quality of the video/images, there appeared to be no issues, with high levels of satisfaction (51.1%). Finally, in terms of the overall platform performance, we have high levels of satisfaction (56.2%).

From these results, it would appear that most of the available virtual learning platforms are delivering of their expectations in the time of COVID-19.

When asked whether the use of online distance learning was better than being at school in person,

8. Would you like to keep using the online distance learning tools even after the pandemic?
在新冠病毒大流行後，請問你還願意繼續使用線上遠距教學嗎

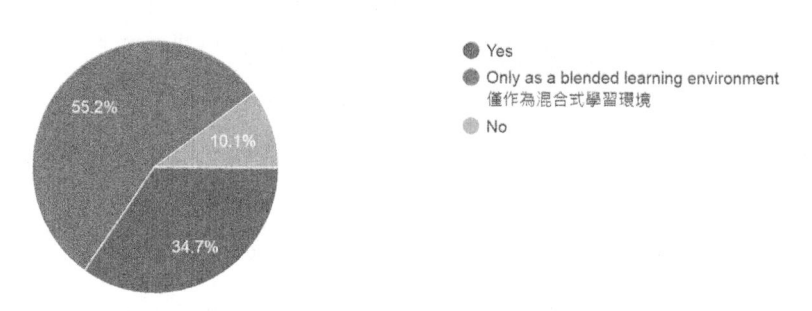

Figure 3. Preference for keeping online distance learning tools post-pandemic from the authors' survey.

a small majority stated that it was only 'Sometimes' (60.6%), with 'Yes' (14.9%) and 'No' (24.5%).

When asked whether they found the use of online distance learning effective for learning, 57.9% said 'Sometimes', with 24.8% 'No' and 17.3% 'Yes'.

The majority of the participants (60.9%), thought that the digital platform provided was both efficient and easy to use. Overwhelmingly, it was also found to be efficient when students were required to upload assignments (84.9%).

When asked whether they thought that gaining knowledge online was the same as in person, most thought that it was, but only by a small margin (55.3%). This result was somewhat surprising. When asked whether the participants preferred the traditional face to face teaching style when compared to distance learning, a fairly high proportion (47.5%) answered 'Sometimes', with 48% saying 'Yes', and only 4.5% answering 'No'.

The final question asked whether the participants would like to keep using the online distance learning tools even after the pandemic was over. Over half (55.2%) stated that only if this was part of a blended learning environment, with 34.7% answering 'Yes', and only 10.1% 'No'.

4 CONCLUSION

Through the use of the online survey, the researchers had gathered valuable information from the participants whose background was mainly from university undergraduate and graduate studies, about how they felt they had been affected by, and adapted to, the COVID-19 situation, together with the extent of their worry surrounding contracting the virus, both for themselves and their family members.

Based on the results from this research, the participants (60%) regarded their mental health had not been affected by C-19. A large majority of them (70%) adapted to the government regulations in terms of isolation and social distancing well. However, most of them (68%) did worry about themselves contracting C-19, and they (81%) did worry about their friends and family developing C-19. Only 8% of the participants had less than four hours sleep during the night. The use of online streaming entertainment and physical exercise are the most popular approaches to alleviate insomnia. More than two thirds of the participants did feel anxiety and loneliness, and half of them suffered a certain degree of depression, as confirmed by other research (Chen, Liang et al. 2020, Hasan and Bao 2020, Husky, Kovess-Masfety et al. 2020).

In terms of satisfaction with university e-learning interaction, the majority of the participants were pleased with its ambience and teaching performance. The majority of the participants found that distance learning is efficient and easy to use. This could be explained by the fact that e-learning has been the mandatory component of all educational institutions, such as colleges and universities in Taiwan for many years. It is interesting to note that slightly over half of them regarded obtaining knowledge online was the same as in person, and they preferred the normal face to face teaching style, as confirmed by (Aguilera-Hermida 2020). Over half of the participants preferred a blended learning environment, but not a pure distance learning approach, post-pandemic.

Compared with 193 other countries, Taiwan has a relatively low number of confirmed cases and deaths. To date, the vaccination rate in Taiwan is less than 1% (CDC 2021). This can be explained by the fact that Taiwanese people have not, so far, suffered much by the pandemic, and feel less inclined to get vaccinated due to the negative side effects of the vaccination in the press and on Social Media. This could change quickly with rising cases, as per the Indian experience of recent weeks. The availability of the vaccine is another factor in this equation.

Based on the survey data, it could be inferred that C-19 has increased the participants' depression, anxiety and loneliness, reduced well-being, and reduced sleep quality. This research also confirmed that the current online learning system in Taiwan was sufficient, in terms of functionality amongst students during the pandemic.

This research has presented empirical evidence on the effects of the C-19 pandemic on youth's mental health and satisfaction of effectiveness of online learning in higher education from students' perspective in Taiwan. Future work will embark on developing strategies for mental health resilience during this and the next epidemics, and the link to Post-Traumatic Stress Disorder (PTSD).

REFERENCES

Aguilera-Hermida, A. P. (2020). "College students' use and acceptance of emergency online learning due to Covid–19." International Journal of Educational Research Open 1: 100011.

Ahmed, M. Z., O. Ahmed, Z. Aibao, S. Hanbin, L. Siyu and A. Ahmad (2020). "Epidemic of COVID-19 in China and associated psychological problems." 51: 102092.

CDC, T. (2021). "COVID-19 Data and Statistics." Retrieved May 12, 2021, from https://www.cdc.gov.tw/En.

CDC, T. (2021). COVID-19 Vaccination Statistics. Taipei, Taiwan.

Chen, R.-N., S.-W. Liang, Y. Peng, X.-G. Li, J.-B. Chen, S.-Y. Tang and J.-B. Zhao (2020). "Mental health status and change in living rhythms among college students in China during the COVID-19 pandemic: A large-scale survey." 137: 110219.

Johns Hopkins University Center for Systems Science and Engineering (JHU CSSE) COVID-19 Data (2021). "Taiwan New cases and deaths." Retrieved 14 May, 2021, from https://g.co/kgs/cyrupK.

Evans, S., E. Alkan, J. K. Bhangoo, H. Tenenbaum and T. Ng-Knight (2021). "Effects of the COVID-19 lockdown on mental health, wellbeing, sleep, and alcohol use in a UK student sample." Psychiatry Research 298: 113819.

Evans, S. L. and R. Norbury (2021). "Associations between diurnal preference, impulsivity and substance use in a young-adult student sample." Chronobiology International 38(1): 79–89.

Hasan, N. and Y. Bao (2020). "Impact of "e-Learning crack-up" perception on psychological distress among college students during COVID-19 pandemic: A mediating role of "fear of academic year loss"." Children and Youth Services Review 118: 105355.

Huang, Y. and N. Zhao (2020). "Generalized anxiety disorder, depressive symptoms and sleep quality during COVID-19 outbreak in China: a web-based cross-sectional survey." Psychiatry Research 288: 112954.

Husky, M. M., V. Kovess-Masfety and Joel D. Swendsen (2020). "Stress and anxiety among university students in France during Covid-19 mandatory confinement." Comprehensive Psychiatry 102: 152191.

Janiri, D., A. Carfì, G. D. Kotzalidis, R. Bernabei, F. Landi, G. Sani, and Gemelli Against COVID-19 Post-Acute Care Study Group (2021). "Posttraumatic stress disorder in patients after severe COVID-19 infection." JAMA Psychiatry 78(5): 567–569.

Norbury, R. and S. Evans (2019). "Time to think: Subjective sleep quality, trait anxiety and university start time." Psychiatry Research 271: 214–219.

World Health Organization (2021). "WHO Coronavirus (COVID-19) Dashboard." Retrieved May 13, 2021, from https://covid19.who.int/.

UNICEF. (2021). "The impact of COVID-19 on the mental health of adolescents and youth." Retrieved May 10, 2021, from https://www.unicef.org/lac/en/impact-covid-19-mental-health-adolescents-and-youth.

Wikipedia. (2021). "COVID-19 pandemic by country and territory." Retrieved 14 May, 2021, from https://en.wikipedia.org/wiki/COVID-19_pandemic_by_country_and_territory
https://en.wikipedia.org/wiki/COVID-19_pandemic_by_country_and_territory#/media/File:COVID-19_Outbreak_World_Map_Total_Deaths_per_Capita.svg.

A CNN-based model for flexible flat cable defect detection

Wei-Chien Hsiao & Shie-Jue Lee*
Department of Electrical Engineering, National Sun Yat-sen University, Kaohsiung, Taiwan

Yen-Ju Hsieh, Ming-Hsuan Lee & Wei-Lin Liao
Department of Information Technology, Cvilux Corporation, New Taipei City, Taiwan

ABSTRACT: Automated optical inspection (AOI) is an important technique for use in the manufacturing industry. Traditional AOI techniques cannot solve certain types of industrial inspection problems. In this paper, we focus on a problem, flexible flat cable (FFC) defect detection, which is difficult to solve using traditional AOI techniques. The goal is to distinguish between normal and defect samples based on rectangular protrusions in FFC images of 256×256 pixels in size. A convolutional neural network (CNN)-based model, with two convolutional layers and three fully connected layers, is proposed to solve the problem. A rectified linear unit is adopted as the activation function and max pooling is adopted for dimensionality reduction in the model. In data preprocessing, we use normalization and appropriate data enhancement techniques to enhance the ability of distinction of the images in the dataset, thereby improving the performance of the model. Experiments show that the CNN model can achieve 96.75% accuracy in FFC defect detection.

1 INTRODUCTION

In the field of computer vision-based defect detection [1], there are many different algorithms available to accomplish the task of automated optical inspection (AOI), i.e., defect detection. AOI [2] is the integration of mechanics, optics, electronic control, and algorithms to replace manual defect detection. In the manufacturing industry, AOI can be used to improve the efficiency and quality of products based on the signals received from the optical processing units. The AOI system relies on specific algorithms to inspect the defects. Besides, the AOI system can also measure the actual length or size of a product precisely beyond the limitations of the human eyes. It has an important role to play in Industry 4.0 [3], thanks to the advances in computer vision technology.

Various machine learning techniques can be implemented in AOI, such as recurrent neural networks to detect wheel defects [4], support vector machines to extract multidimensional visual features [5], and convolutional neural networks (CNNs) for surface defect detection ([6], [11], [12], [13], [14]). The problem we try to solve in this study is to determine whether two rectangular protrusions are within a certain ratio, as shown in Fig. 1. This is a very common application in the field of industrial inspection and can be fulfilled using deep neural networks.

Our dataset is a collection of images. Each image includes a part of flexible flat cable (FFC) with rectangular protrusions on both sides of the board, as shown in Fig. 2. The goal is to determine whether their areas and shapes can be regarded as the same.

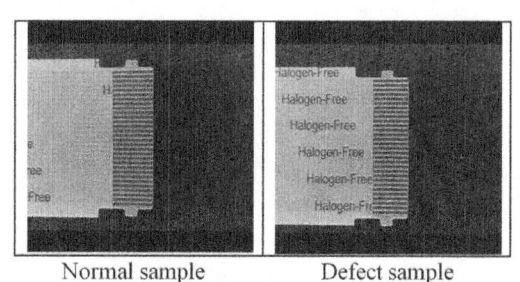

Normal sample   Defect sample

Figure 1. Two different classes in the dataset.

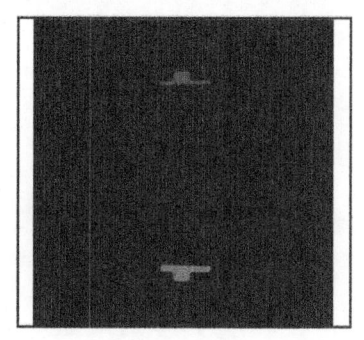

Figure 2. The compare part of an image.

*Corresponding Author

DOI 10.1201/9781003278474-11

If in a sample the size difference between two rectangular protrusions on the board exceeds a certain value, the sample is considered to be defective; otherwise, the sample is considered to be normal. It is difficult for human eyes to detect the difference; only 80% detection accuracy could be achieved through manual inspection.

One possible alternative to manual detection is through image segmentation. Image segmentation entails manually labeling each image sample and creating a mask for each image in the training set (the mask can be a rectangle, a polygon, or any other shape). Examples of image segmentation models include U-Net [7], Mask R-CNN [8], Fast R-CNN [9], and Yolact [10], among others. The image segmentation model is trained to recognize a specific shape and create a mask to conform to that specific shape. To perform image segmentation on an image, first, a mask should be manually marked in the image, and then through proper training, the model can produce an adequately fitting mask for the image. Finally, defect detection is done by determining whether or not the mask generated from the model is within an acceptable range. However, this approach requires a lot of manual intervention, consuming a lot of time and labor in preprocessing the images.

We propose a method to overcome the aforementioned defect detection constraints in a simpler way. We concentrate on the comparison of the target areas by extracting relevant features to recognize a specific shape without creating a mask. It saves a significant amount of time invested in labeling masks, and only a binary classification is provided. We do not need to determine the shape or range through using a complex model, if the purpose of an inspection is to distinguish a normal sample from a defective sample. Our method uses a simple CNN model to extract the features directly from the images without any manual intervention. Our method can reduce the cost in terms of time and labor, an important KPI (key performance indicator) used in this industry.

The remainder of this paper is organized as follows. Section 2 describes the dataset we are working on. Section 3 discusses the proposed defect-detection model framework and all relevant details. Section 4 presents experimental results. Finally, Section 5 concludes the paper.

2 DATASET

The dataset is provided by a private connector manufacturing enterprise and contains two classes of samples, normal and defect, as shown in Table 1.

Each image sample is part of an FFC with a different watermark, as shown in Fig. 1. The goal is to distinguish between normal and defect samples based on the two rectangular protrusions shown in Fig. 2.

As shown in Table 1, the dataset contains a total of 402 samples, of which 192 are normal and 210 are

Table 1. Dataset.

Class No.	Class	Count	Description
0	Normal	192	The ratio of protrusions is in an allowable range of interval
1	Defect	210	The ratio of protrusions is not in an allowable range of interval

defective. Clearly, the classes are balanced and comply with the requirements for building a classification model. The criterion for distinguishing between normal and defect samples is whether the ratio of the two rectangular protrusions is within an allowable range of interval. If the ratio is outside the allowable range of interval, the sample is considered to be defective and placed in the defect sample class.

3 THE PROPOSED METHODOLOGY AND MODEL

The architecture we propose is shown in Fig. 3. The images in this model were resized to 256×256 pixels for faster computation without compromising the detection quality. The data augmentation technique is implemented by randomly flipping sample images in the dataset, with a 50% chance of vertical flip, and normalizing the sample by 3 channels (mean and SD), which are [0.2414, 0.2301, 0.2180] and [0.2839, 0.2661, 0.2516], respectively. This helps the model focus on the rectangular protrusion areas in the samples, and also enhances the size and quality of the training dataset.

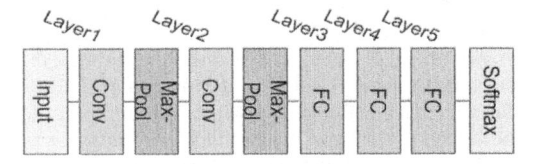

Figure 3. The architecture of our model.

On the other hand, we used five-fold cross-validation to divide the normal samples and defect samples from the dataset into five classes, which, in turn, form the training set (80%) and validation set (20%), as shown in Fig. 4.

The loss function we adopt is cross-entropy lloss:

$$loss(x, class) = weight[class](-x[class] + \log(\Sigma_j(x[j]))) \tag{1}$$

$$loss = \frac{\Sigma_{i=1}^{N} loss(i, class[i])}{\Sigma_{i=1}^{N} weight(i, class[i])} \tag{2}$$

Note that we did not use mean square error (MSE).

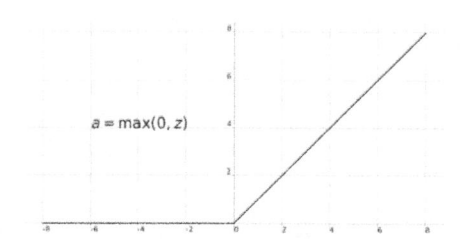

Figure 4. Five-fold cross-validation.

since this is for the data generated from a normal distribution. For two categories, the data are more reasonably generated from a Bernoulli distribution, for which cross-entropy is more suitable.

The activation function we adopted is a rectified linear unit (ReLU), which is shown in Fig. 5. ReLU helps to prevent the exponential growth and solve the gradient problem. Besides, less computing time is involved in training the model.

$$a = \max(0, z)$$

Figure 5. Activation function ReLU.

Finally, we adopted the stochastic gradient descent (SGD) with momentum to train the model. Momentum is a method that helps accelerate SGD to converge in the dampened oscillations in relevant directions, as shown in Fig. 6. Based on SGD frequent updates, the steps taken toward achieving the minima of the loss function involve oscillations, which can help come out of the local minima of the loss function. Note that by early stopping, the training will stop when the desired training accuracy is hit seven times consecutively.

SGD- without momentum   SGD- with momentum

Figure 6. Comparison between SGD and SGD-with-momentum.

The model is trained in the open-source machine learning framework PyTorch with hardware NVIDIA GPU RTX-2060, using Deep Learning SDK. We trained our CNN model on the GPU, and NVIDIA

Tensor Cores was used to accelerate the training time from CPU computing to GPU computing.

4 EXPERIMENTAL RESULTS

The 5-fold cross-validation results are shown in Table 2, with an average accuracy of 96.75%, which is better than the accuracy of 80% achieved by manual detection. The accuracies of training and testing at each epoch are shown in Fig. 7, and the cross-entropy losses are shown in Fig. 8. As can be seen, the model learned well and overfitting did not occur.

Table 2. Results of 5-fold cross validation.

Fold No.	Accuracy	Stop Epoch
1	97.5%	98
2	98.75%	82
3	97.5%	77
4	93.75%	73
5	96.25%	96
5-Fold standard deviation	1.2748	
5-Fold average	96.75%	
Recall	0.9459	
Precision	0.9722	
F1-score	0.9589	

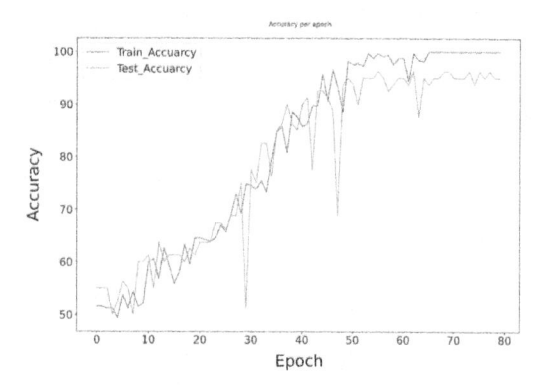

Figure 7. Accuracies at each epoch.

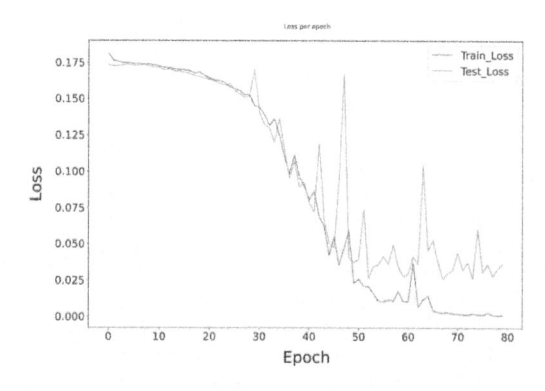

Figure 8. Loss of training and test per epoch.

5 CONCLUSION

Our proposed model has a few advantages. First, it reduces the time required for data training and labeling. Second, it has high accuracy without overfitting. Third, it is the first CNN application in FFC. In the future, the model can be deployed in a production line for edge computing and real-time anomaly detection. This study may encourage the application of machine learning to automatic anomaly detection and help push Industry 4.0 forward.

ACKNOWLEDGMENT

This work was supported by the , Ministry of Science and Technology through grant MOST-108-2221-E-110-046-MY2, the NSYSU-KMU Joint Research Project (#NSYSUKMU 110-KN002), and the "Intelligent Electronic Commerce Research Center" from the Featured Areas Research Center Program within the framework of the Higher Education Sprout Project funded by the Ministry of Education in Taiwan.

REFERENCES

[1] Aleem, S., Capretz, L. F., & Ahmed, F. (2015). Benchmarking machine learning technologies for software defect detection. arXiv preprint arXiv:1506.07563.

[2] Azizah, L. M. R., Umayah, S. F., Riyadi, S., Damarjati, C., & Utama, N. A. (2017, November). Deep learning implementation using convolutional neural network in mangosteen surface defect detection. In 2017 7th IEEE International Conference on Control System, Computing and Engineering (ICCSCE) (pp. 242–246). IEEE.

[3] Bolya, D., Zhou, C., Xiao, F., & Lee, Y. J. (2019). Yolact: Real-time instance segmentation. In Proceedings of the IEEE/CVF International Conference on Computer Vision (pp. 9157–9166).

[4] Garcia, G. R., Michau, G., Ducoffe, M., Gupta, J. S., & Fink, O. (2020). Time Series to Images: Monitoring the Condition of Industrial Assets with Deep Learning Image Processing Algorithms. arXiv preprint arXiv:2005.07031.

[5] Girshick, R. (2015). Fast r-cnn. In Proceedings of the IEEE international conference on computer vision (pp. 1440–1448).

[6] Gobert, C., Reutzel, E. W., Petrich, J., Nassar, A. R., & Phoha, S. (2018). Application of supervised machine learning for defect detection during metallic powder bed fusion additive manufacturing using high resolution imaging. Additive Manufacturing, 21, 517–528.

[7] He, K., Gkioxari, G., Dollár, P., & Girshick, R. (2017). Mask r-cnn. In Proceedings of the IEEE international conference on computer vision (pp. 2961–2969).

[8] Jia, H., Murphey, Y. L., Shi, J., & Chang, T. S. (2004, August). An intelligent real-time vision system for surface defect detection. In Proceedings of the 17th International Conference on Pattern Recognition, 2004. ICPR 2004. (Vol. 3, pp. 239–242). IEEE.

[9] Kumar, A. (2008). Computer-vision-based fabric defect detection: A survey. IEEE transactions on industrial electronics, 55(1), 348–363.

[10] Lasi, H., Fettke, P., Kemper, H. G., Feld, T., & Hoffmann, M. (2014). Industry 4.0. Business & information systems engineering, 6(4), 239–242.

[11] Liao, H. C., Lim, Z. Y., Hu, Y. X., & Tseng, H. W. (2018, July). Guidelines of automated optical inspection (AOI) system development. In 2018 IEEE 3rd International Conference on Signal and Image Processing (ICSIP) (pp. 362–366). IEEE.

[12] Park, J. K., Kwon, B. K., Park, J. H., & Kang, D. J. (2016). Machine learning-based imaging system for surface defect inspection. International Journal of Precision Engineering and Manufacturing-Green Technology, 3(3), 303–310.

[13] Ronneberger, O., Fischer, P., & Brox, T. (2015, October). U net: Convolutional networks for biomedical image segmentation. In International Conference on Medical image computing and computer-assisted intervention (pp. 234–241). Springer, Cham.

System Innovation in a Post-Pandemic World – Kin-Tak Lam et al. (Eds)
© 2022 Copyright the Author(s), ISBN: 978-1-032-24392-4

Research on the application of virtual reality in the education and learning of panoramic photography

Chien-Yu Kuo*

Department of Education, National University of Tainan, West Central District, Tainan City, Taiwan, ROC

ABSTRACT: In this project, an interactive guidance interface is built in the virtual environment presented by virtual reality (VR) glasses to explore whether this interface is helpful for positioning and prompting when viewing VR images. It is explored in the VR environment whether the usability of the interactive guidance interface and the actual guidance effect are in line with the expectations.

The goal of the VR project developed in this paper is to provide elementary school students a direct and interesting experience of visiting the National University of Tainan. The target users of this project are elementary school students. The target users are able to use mobile devices. The produced VR has two characteristics: interaction and immersion.

Keywords: Panoramic photography, virtual reality, education and learning, interactive guidance interface.

1 INTRODUCTION

The advancement of technology has brought convenience to life, enriched our life experience, and made life more fun. Improving education has been the focus of several efforts around the world. Over the past couple of decades, the usage of digital devices and other interactive media applications such as interactive books and video games has been added to the list of learning tools in pre-school classrooms (Romano et al., 2007). Currently, in the 21st century, learning is being highly affected by emergent cultures. Attitudes of learners have shifted toward self-paced learning, openness, and teamwork from the traditional learning method in class learning. Accordingly, new technologies have to be combined with traditional methods to attract learners (Tang et al., 2009; Connolly et al., 2009). Therefore, making use of the potential of technology in the learning process may stimulate learners' attention, ease their test anxiety, provide them with a rich learning experience, and so on. For example, animation with animated pictures and sounds may attract learners' attention. The interactive responsive systems, such as Kahoot, turn the printed test into e-format, making the assessment process more interesting and providing the learners with instant feedback. Furthermore, using technology, such as virtual reality (VR), they may observe the phenomena which they are not able to touch and visit. This kind of experience may make "knowledge presented in text formats"

more vivid. The key to making the above-mentioned potentials happen is controlled by the classroom teachers. They must observe the potentials of technology for learning enhancement, be willing to learn the technology, and cultivate their abilities to integrate technology into teaching and learning. Teacher education is critical to developing a teacher's technological pedagogical content knowledge, of which technology knowledge is one core component. This paper focused on one emerging technology, VR, and discussed its potential for teaching and learning. VR uses computer analogy to generate a virtual world in a 3D space, providing users with an analogy about vision and other sensory, making users feel as if they are immersed in the environment, and can observe things in 3D space in real time without limitation. Compared with the traditional lecture- and listen-based teaching and learning method, VR technology may have students interact with the knowledge more comprehensively without physical distance limitation. VR presents phenomena or knowledge in 3D formats via mobile devices. Learners can observe complex phenomena or knowledge shown in the real world.

Furthermore, learners may interact with the objects to get explanations or instant feedback. For example, learners could learn history, geography or more subjects by using VR. Compared with the text in the book, the photos presented in VR allow students to truly experience the difference in terrain, climate, and environment created by them. Using VR to present the history, learners are able to visit historical sites. Medical students could simulate surgery through VR. Using

*Corresponding Author

 DOI 10.1201/9781003278474-12

VR to learn the beauty of arts, students, wearing a Google Cardboard glass, can feel like walking into an art gallery to actually appreciate Van Gogh's picture. Grubbs pointed out that under certain situations, VR also provides a safe learning environment.

Grubbs said "Students can choose various tubes by themselves now, fit them together, and see what's going on. As a result, if an explosion occurs, it will only explode in the virtual world, and no one will be injured" (Wind Media, 2017). In addition, VR teaching makes the abstract content intuitive, and visually displays some processes that are difficult to achieve and observe under ordinary conditions, which is convenient for observation and understanding, and is conducive to learning and mastering teaching materials. Make the jerky and unfamiliar knowledge visualized, which stimulates students' interest in learning and mobilizes their enthusiasm for active learning (Lin Greta, 2020).

Another advantage is that VR makes teaching simplified, increases the amount of information, improves the teaching speed, and saves time, which not only effectively expands the class time capacity, but also reduces the teacher's labor (Luzhao Mao, 2021). Lastly, VR can invigorate the classroom atmosphere, deepen and consolidate the teaching content, so that students feel the joy of learning and enjoy learning. Therefore, the knowledge expression ability of the class is stronger, and the students will be more impressed (Zixuan Li, 2019). With the development of science and technology, this kind of technology will be combined with network technology, simulation technology, artificial technology, and other high-tech applications in education and teaching, which will be the trend in the teaching scene in the future. After pre-service teachers observe the potential of VR, the way to teach them how to make use of VR is critical.

The pre-service teacher has to create the VR based on their knowledge and interest in classroom teaching. That is, having pre-service teachers develop a VR, which they may use in teaching in the future may have them observe the task value of conducting a VR product. Such a project experience will help them to transfer into their teaching contexts when becoming an in-service teacher. Therefore, this study aimed to demonstrate a process of creating a VR project multisensory stimulation, so that users feel as if they are in a real environment. In the interactive VR design, users can directly control and operate the scenes and things in the VR, and give them quick feedback via computer calculations. The response to provide simulated real territory interaction and response. Nonetheless, owing to the limitations of the growth of information technology, though the current VR can offer users realistic situations, it still cannot build the same environment like the real world. Hence, in the application design of VR, a large proportion of them have special themes and design significance to induce the imagination of users, and then match the created VR scenes to build a more tactic VR application (Chen Zhiming, 2012).

2 PANORAMIC PHOTOGRAPHY

Panoramic mode is also known as panoramic stitching and has other names too. Panoramic mode is a distinctive function designed for shooting landscapes. As long as the user splits and photographs the view in the same horizontal method, some models can be vertical, even four-square or nine-square stitching. You can use the panoramic function of the camera to collage the photos one by one into a long filled landscape. This mode will automatically keep the 15% overlap of the final shot for collages, and the camera will save it. Let the images shot here have the same exposure value, and then use the bundled picture synthesis software to adapt and fine-tune the overlapping portions to achieve a tactic connection (NASIM MANSUROV, 2019) (Figures 1–4).

3 DEVELOPMENT OF A VR PROJECT

The first stage of developing a VR project is to establish the goal and the target users and analyze the needs and experience of the users. The goal of the VR project developed in this paper is to provide elementary school students a direct and interesting experience of visiting the National University of Tainan. The target users of this project are elementary school students. The target users are able to use mobile devices, which is essential for watching VR. The project not only hopes to make teaching more interesting, but also to make learning highly related to the real world. The students could wear Google Cardboard glasses to watch the VR in 360 degrees. The equipment needed to watch VR includes a mobile device with an Android system and Google Cardboard glasses. The second stage is to design the story, including each scenario and information that will be embedded in the VR. The VR project designed is composed of five pictures of the National University of Tainan. The entry scenario is the "Red House," the first building visitors see when walking into the National University of Tainan. The corridor in the center is called Time and Space Corridor. The second picture is of the famous "Clock Tower" at the school. Every year at the graduation ceremony, one famous teacher hosts the bell ringing ceremony to represent the graduates and wish them great prosperity. The third photo is of the statue of Xin Chuan between the Red House and Cheng Zheng Building of our school. It was built by alumni when the Cheng Zheng Building was completed that year. In addition to condensing the relationship between the students and the school, it represents the spirit of "inheriting the past." The fourth picture is of the playground of our school, which allows many students from the physical education department to practice their professionals or nearby residents to do exercise. The fifth picture is of the two dormitories of our school. They are a dormitory for first-year girls and a mixed dormitory for boys and girls. The third stage is to produce materials to be used in the VR. Several steps to produce materials are described as follows.

Figure 1. The Red House.

Figure 2. Clock Tower.

Figure 3. Xin Chuan.

Figure 4. The playground.

Figure 5. Dormitories.

First, five shooting spots of the National University of Tainan were selected. Second, the panoramic camera was used to take 360-degree pictures (see Figures 1–5). Third, these photos are input into the Unity program to make the VR. Fourth, the produced VR is tested in an Android-based mobile phone.

4 RESULT – THE VR PROJECT

I hope to present not only new knowledge, but also to cultivate their ability to find knowledge. With my VR teaching materials, elementary school students can explore various skills. The teaching method of my VR project is different from the traditional education in the past. It has been designed to act as a platform for converting traditional books into interactive books that use Augmented Reality (AR) concepts. It can present the flat content in a 3D scene. By watching a virtual environment, people can feel like they are in the real environment. The produced VR has two characteristics:

1. Interaction: In addition to the presentation of the simulated scene, the user can interact with the virtual scene objects. By turning the direction to find the button on the screen, some explanation messages can be jumped out to produce a more realistic feeling.
2. Immersion: The virtual scene can satisfy various sensory feelings to help blend into it. The characteristics and interactivity of VR allow us to see different aspects of the world in many products and even in the field of education, not limited to images in textbooks.

5 CONCLUSION

Compared with just viewing some of the history of National Tainan University on the website, users see the VR introduction to Tainan University. Through the display of VR and the graphic guide, users will have a clearer understanding of the university. With the progress of the times, the future teaching mode will gradually change into different methods. VR is an inevitable trend. For teachers and students, it is possible to teach in cities and villages in the future. Teachers in cities can use VR teaching methods, let students experience the appearance of farming, and teachers in the countryside can let the students who rarely go out to see the beauty of the world.

The difficulty of production should be that Unity is not very simple software. It takes a lot of time to learn to produce. Not all teachers like programming, so there will be certain difficulties in implementation.

If there are teachers and students who are willing to learn this aspect of technology in the future, it is recommended to start with the production of other programs, simple and deep, and then learn AR & VR production after a period of time to be handier.

REFERENCES

Nasim Mansurov (2019), Panoramic Photography Tutorial
Romano et al., 2007 Luzhao Mao (2021),VR distance learning, the last mile of fluorescent digital learning Lin Greta (2020), Medical VR
Tang et al., 2009; Connolly et al., 2009
Zhi-Ming Chen (2012), virtual reality
Zi-Xuan Li (2019) 2019, Educational Innovation 100 Joe Bardi (Published: 03/26/2019 | Updated: 09/21/2020) What is Virtual Reality? Definition and Examples

System Innovation in a Post-Pandemic World – Kin-Tak Lam et al. (Eds)
© 2022 Copyright the Author(s), ISBN: 978-1-032-24392-4

A study on the model of township brand design upon placemaking—Beidou Town

Shu-Huei Wang*
Department of Digital Design, University of MingDao, Pitou, ChangHua, Taiwan

ABSTRACT: The administration of Taiwan announced the year 2019 as the first year of placemaking and implemented the policy of local revitalization for the next two years. The actual results of the implementation of the policy in the local governments, public organizations, and schools as well as the differences between the central and local administrations are to be investigated. Brand establishment in the placemaking strategy would be the scope of study in this research. A case study on the Beidou town and a questionnaire survey were conducted for an understanding of a model for local creation through the practice of township brand design.

1 INTRODUCTION

Shinzō Abe, the prime minister of Japan, executed the policy of placemaking in 2014 with a view to solve the problems of decreasing employment opportunities in towns and the countryside and the huge gap between cities and towns due to the aging local towns and outflowing businesses. The example of Japan was borrowed for the implementation of the local revitalization policy in Taiwan. Local industries needed to be developed based on regional characteristics for attracting young people to their hometowns for work, which not only solved the problem of uneven distribution of population, but also rendered a balanced Taiwan. The Executive Yuan approved the National Tactical Project of Placemaking on January 3, 2019 and announced the year of 2019 as the first year for local revitalization. Now that it has been almost two years since the implementation of the policy of local revitalization, the difference between the expected and the actual results needs to be investigated. Ex- changes among cities and twin towns were quite popular in early days in the global community. Various business activities were held based on the characteristics of the city, which was exactly the idea behind establishing a city brand. Also, the government executed the policy of one village with one feature hoping that township characteristics could be built and economic and commercial development might be promoted.

2 METHODOLOGY

The policy of placemaking was implemented in 2019 with the aim of boosting economic and population growth in small towns for a balanced Taiwan. A brand established by placemaking was under study in this research, and the Beidou town was targeted for developing it into a new township brand model in the post-pandemic era. It was expected that resources could be integrated for sustainable operation through such a brand, which could be further marketed and extended to the global community via technology.

This research was based on literature review and case studies to construct a brand design model for placemaking via the Flower model and theory.

3 LITERATURE REVIEW

3.1 Placemaking

The placemaking policy of Japan with three strategies was borrowed for the implementation of local revitalization in Taiwan in 2019. The first strategy was the Regional Economy Society Analyzing System (RESAS) followed by talent support via the placemaking talent cultivation system and financial support through placemaking delivered funds. The National Strategic Plan for Placemaking was proposed for the organization structure. The Executive Yuan set up the Placemaking Journal in coordination with the central departments, local governments, people in charge of private industries, scholars, and experts, and the National Development Council acted as a platform and was responsible for the integration, planning, and matching of placemaking proposals. Township (city, district) offices, local industrial and academic institutions, and associations worked together to draw a placemaking project considering regional DNA and resources (National Development Council, 2017).

The five strategies adopted for placemaking included business investments in hometowns, introduction of technology, joint participation of industry, government, and academic associations, integration

*Corresponding Author

DOI 10.1201/9781003278474-13

and distribution of placemaking resources, and brand building and global marketing. Based on the principle of industrialization of culture and making industry cultural, culture is a national soft power and design is an underlying force behind industrial transformation. Local uniqueness and core values should be ensured before bringing in creativity (design), innovation (productivity), and entrepreneurship (marketing). A total of 134 towns with high population flight/exodus and economically weak residents were shortlisted as priority regions for placemaking in compliance with the rate of natural increase in population size and residents' incomes (National Development Council, 2019).

3.2 City brand versus township brand

The idea of the creative city proposed by Landry (2000, 2008) was to rejuvenate and renew a city. Reconstruction of a city via creative settlements to achieve the dual goal of a renewed city and a reprosperous economy is one of the strategies adopted in cultural and creative policies in many countries. Landry (2008, pp. 309-318) presented a cycle of city creativity to evaluate creative proposals on different development stages, including (1) assisting people in generating concepts and projects; (2) realizing concepts; (3) establishing networks for concepts and projects and increasing circulation and marketing; (4) putting into place mechanisms like cheap spaces for rent, and incubating organizations or opportunities for display and exhibition; and (5) popularizing the results in the city, establishing a market and user groups, and motivating new ideas through reviews and assessments. He also indicated that cultural uniqueness resulted in differences between cities. Putnam (2000) stated that a prosperous society requires healthy civic awareness; on the contrary, social capital and trust decreased. Florida (2003; 2006) considered making a creative city with 3T (talent, tolerance, and technology) might lead to economic growth. After the United Nations Educational, Scientific and Cultural Organization (UNESCO) approved The Creative Cities Networks in 2004, all cities could learn and share experiences, innovate their own cultures, promote the creative industry, and develop urban economy through this platform.

In contrast to a city brand, a township brand was rarely mentioned and replaced by a business brand mostly. For instance, we might associate Toyota with the Toyota City in Aichi Prefecture, Japan, easily because the characteristics of a certain industry could lead us to a certain town. However, towns have their own charisma just like cities do. Therefore, towns should set up their unique township brands as well. As far as the functions of a regional brand are concerned, Jun-xian Wu and Yang, Ying-xian (2009) gave a good definition: The connections of regional products, branded service, local symbols or totem, and brands generated a good cycle for production and marketing, which could further bring in capital and talents outside the region to achieve the aim of activating regional economy continually. There was an indigenous DNA, history, status quo and vision for

the future in every place. The diversified integration of talents and regional resources might motivate industrial development and local culture to set up a local brand (Ya-ping Guo, 2018). Brand building, one of the five strategies for placemaking, is the most efficient means of integrating township brands with local resources. The policy of one town with one characteristic proposed earlier had no horizontal connection at the moment.

3.3 Cultural and creative settlements versus vacant spaces

Combination of the local industry and the settlement preservation initiatives might spur regional development (Ye, Zheng-yang, 2010). It was important to consider geographical factors in establishing creative settlements. For the smooth functioning of the Cultural and economic strategies adopted for cities, they should be monitored and updated from time to time through common understanding and power-sharing so that a new local life with a new economic style and competitiveness could be developed (Gu, Jia-yu, 2010).

According to Frey (2009), a creative environment plays a critical role in the upliftment of masses, and the information related to the organization of a creative place, interpersonal network, and work and life models could be obtained from the present culture and economy or from the fundamental theories for creative development. Culture is a continually evolving and accumulating aspect of human settlements and communities. A cultural change does not change the structure of a building, but it changes the social structure of a community. Nevertheless, some cultural changes might bring a change to both at the same time! Creative talents, industries, and networks come together due to the availability of new space, policy, environment, and people in the creative settlements; however, managing different industries, marketing, and brand building are some of the challenges facing creative settlements. Satisfy consumers through enhancing their experience of living in a township or building city or township brands for sustainability was an issue under consideration.

4 A STUDY ON THE INFLUENCE OF TOWNSHIP BRAND DESIGN ON PLACEMAKING

The placemaking policy in Taiwan was based on creativity (design), innovation (productivity), and entrepreneurship (marketing) to create industrial charisma and values. The case of the Beidou town for placemaking revealed the strength of design. Umebara Makoto from Kōchi Prefecture, Japan, was responsible for local design, which is different from general commercial design. He studied in Spain at the age of 25, then quit his job to finish his trip across the United States, and during this trip, he stayed in San Francisco for a short period of time when he was 29. He established Umebara Design Office, Kōchi Prefecture, after returning to his hometown in 1980. His idea of industry × design = landscape was manifested in developing industries through design. Take, for example,

"Bonitos Fished by Handline," a big win that created billions of Japanese yen. The absolute value of the locality was found out by the design concept to reveal the peculiar landscape of the area. The author interpreted the landscape via visualized design, manifesting the local story and its characteristic values. Mr. Umebara Makoto has created numerous local specialties over the past 40 years, and he was a real maestro in turning the hopeless end-of-rope products into valuable products through his turnaround design techniques. The local bonito product was characterized by Tosa, Handline Fishing, and Overburning straw. This special fishing approach created an identity for Kōchi followed by much appreciation. An in-depth understanding of the values generated during place-making through design practice was achieved through investigating these local designs against their special meanings (News & Market, 2021).

4.1 Geographical location of Beidou

Beidou is located on the southeastern side of Changhua County, north of Dongluo River (formerly Zhuoshui River) with a total area of 19.2547 square kilometers and a population of 33,193 (as of November 2015). It used to be an important river port and a distribution center for merchandise in central Taiwan during the Ching Dynasty since trading with Fukien Province was made through Hsihu and Lukang. The change in the flowing channel of Zhuoshui River was one of the major reasons for the change and developments in the settlements in Beidou. As Dongluo River gradually lost its connectivity downstream with Zhuoshui River, hydrolutyte increased and the railway did not pass by it during the era of Japanese occupation, and as a result, its role of the traffic hub was replaced by Yuanlin slowly. The former name of Beidou is Baota because it is on the north side of Dongluo River, and in Chinese, the north of a river is called Yang. Consequently, it was also renamed Lo-yang (Wikipedia, 2015).

4.2 Historical culture of Beidou

The history of Beidou is manifested in the historical buildings on the old street. Row houses with decorated archways are an important feature of the old street in Beidou due to its spatial characteristic for business. The inscription on the street stele says: "...for the north, Tianhou Temple was built thereof facing south; Tudi Shrine was set up on northwest to respect the Earth God and protect people, which fulfilled piety and human care. There were homesteads on both sides, which was called Beiheng Street. The middle and back streets were located from east to west with two big alleys between them. There was a cross street in the south. A number sign (intersecting parallels) shape was established to separate streets from alleys. There were a bamboo fence, a trench and a fence gate to prevent thieves and burglars..." (Corpus of Inscriptions on Tablets in Central Taiwan,

1808). Tianhou Temple was renamed Dian-an Temple. From the old street in Beidou from the Ching Dynasty, the era of Japanese occupation until now, Gongqian Street, Gonghou Street. and Douyuan Road were mainly focused, which are located in a number sign (intersecting parallels) fashion of old times. The buildings on both sides is a reflection of the construction style during the Japanese reign. Xie, Rui-long, a culture worker on Beidou, described the street view as: "Tianhou Matsu Temple was the center and gates were installed on the four corners of the block, which were named as East Gate, West Gate, South Gate and North Gate. The big street in front of the temple (Gongqian Street) led to the ferry boat terminus of Dongluo River directly. Beiheng Street (Douyuan Road) served as the east-west main road with crisscrossing paths forming a number sign (intersecting parallels) shape, which was also the first street shop built by urban planning in Taiwan." (Xie, Rui-long, Hong, 2009). On the other hand, Dongluo River and its branch, Qingshui River, nourish the economy and life in Beidou. "The name of this river came from the cultural settlement in Beidou; at the same time, this cultural settlement of Beidou developed stronger through the nutrients it provided. Therefore, the cul- ture on the streets of Beidou was full of a heavy river port lifestyle. The historical culture and memories of life were closely connected to Dongluo River. It can be said to be the best stage to display the culture of Dongluo River for the past few centuries." (Xie, Rui-long, Hong, Qing-zong and Lin, Jian-cheng, 2005). Beidou inherited the culture of Dongluo River and nurtured Beidou culture.

Lin, Chong-xi (2012) indicated a cultural ecosystem consisting of survival and development efforts when different circumstances were confronted with, which included plenty of microsystems, small, medium, and large. Operation of each microsystem consisted of a variety of interactions among life forms of similar type, life forms of different kinds, and between life forms and the environment. From the perspective of the cultural ecology in which an individual exists, the aforesaid systems consisted of families, interpersonal networks, communities, societies, environments, knowledge skills, values, and faith, among others (Figure 1).

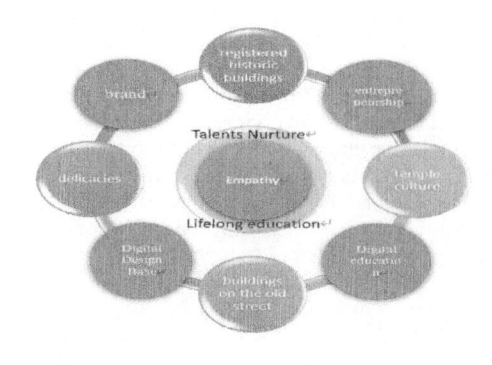

Figure 1. The model construction of Beidou town cultural eco system.

5 CONCLUSION

(1) *Registered historical buildings*: The Cultural Affairs Bureau, Changhua County, had registered several historical buildings as cultural heritage, including houses with Japanese style, senior provincial official residence, Bao-Jia Households Office, the red-brick market, the Far East Theater, and the Earth God Shrine, among others.

(2) *Delicacies*: Fang, Ying-lan mentioned in her master's dissertation (2014) that "1. characteristic snacks and specialties of Beidou had historical context; 2. consumers' opinions about characteristic snacks and specialties of Beidou included a sightseeing tour, a feast of delicious food and a gift for friends and relatives." The most famous delicacies in Beidou are Taiwanese meatballs (Ba Wan), the sandwiches of Hung Rui Chen Bakery, and meat floss (Rousong), among others.

(3) *Temple culture*: Dian-an Temple built by the residents of Jiumeizhuang during Emperor Kangxi's reign in Qing Dynasty was the center. The Queen of Heaven is worshipped in the temple. Goddess Matsu stays here when making her royal progress every year. It has enjoyed a booming pilgrimage, become an important belief center for local residents, and encouraged related economic development. As a result, certain impact on the township brand of Beidou could be made by investigating the history of Dian-an Temple and integrating the cultural and creative industry.

(4) *Buildings on the old street*: For the buildings with a history on the old street of Beidou, some were waste without any management and some could not be acquired due to too many owners. Besides, traffic jams were prevalent when activities were held because the alleys and lanes were too narrow. If the government or relevant organizations could not manage these old buildings, then either private homesteads would be rebuilt or they would be sold, and those historical memories about the old street would fade away slowly with the old generation.

REFERENCES

Frey, O., 2009, Creativity of places as a resource for cultural tourism, PP: 135–154. Springer Netherlands.

Florida Richard, translated by Fu, Zhen-kun, 2006, Cities And The Creative Class. Taipei: Heliopolis Culture Group.

Florida Richard, translated by Zou, Ying-yuan, 2003, The Rise of the Creative Class. Taipei: Baoding Books.

Gu, Jia-yu, 2010. A Study on Creative Settlements in Taipei Blocks – an Example of Shuanglian and Shida Blocks. NCCU Institutional Repository http://nccur.lib.nccu.edu.tw/handle/140.119/52528.

Guo, Ya-ping, 2018, A Study on Strategies of Building Lo- cal Brands in Taiwan. National Taipei University of Technology. An unpublished master's dissertation.

Landry, C., 2000, The Creative City: A Toolkit for Urban Innovators., London: Earthscan.

Landry, C., 2008, The Creative City: A Toolkit for Urban Innovators, P. 179.

Lin, Chong-xi, 2012, The Conservation of Cultural Heritage Needs A New Culture: A Cultural Ecology Viewpoint for the Industrial Heritage Conservation, Journal of Cultural Property Conservation, No.22, pp.73–84

Placemaking by Design Flip Project promoted by National Development Council, 2017, Date of reference: 2020/12/26 https://www.ndc.gov.tw/Content_List.aspx?n=4A000EF83D724A25

Placemaking Policy/Explanation of Placemaking Policy promoted by National Development Council, 2019, Date of reference: 2020/12/26, https://www.ndc.gov.tw/Content_List.aspx?n=78EEEFC1D5A43877&upn=C4DB8C419A82AA5E

Putnam, R., 2000, Bowling Alone: The Collapse and Revival of American Community., New York: Simon and Schuster.

Umebara Makoto, forerunner of local design in Japan. Design for the place: Place of Production Expressed by Design. Date of reference: 2021/01/01. Source of reference: News & Market. https://www.newsmarket.com.tw/blog/120771/

Wu, Jun-xian and Yang, Ying-xian, 2009, Strategies of Innovating Local Industry/Building Local Brand and Developing Local Industry. Taipei: Wu-Nan Culture Enterprise.

Xie, Rui-long, 2009, Beidou Chorography. Changhua: Beidou Township Office, Changhua County. P. IV-1.

Xie, Rui-long, Hong, Qing-zong and Lin, Jian-cheng, 2005, In Love with Beidou. Changhua: Cultural Affairs Bureau, Changhua County. P. 73–74.

Ye, Zheng-yang, 2010, Preservation of Settlement Potential and Evaluation of Activating Proposals – an Example of Sinpi- tou and Qiedongjiao Settlements

Wikipedia. 2015. Date of reference: 2015/12/17. Source of reference: https://zh.wikipedia.org/wiki/%E5%8C%97%E6%96%97%E9%8E%AE_(%E5%8F%B0%E7%81%A3)

System Innovation in a Post-Pandemic World – Kin-Tak Lam et al. (Eds)
© 2022 Copyright the Author(s), ISBN: 978-1-032-24392-4

Application of VR characteristics and narrative methods in artistic creation: A case study of "Inori"

Chiu-Ping Wang & Pai-Ling Chang*
Department of Digital Multimedia Arts, Shih-Hsin University, Taipei, Taiwan

ABSTRACT: The unique environment brought by the immersive virtual space of VR provides artists with new possibilities, new perspectives and creative methods. This is a world beyond the reach of graphic painting. We hope to know whether artists consider the characteristics and narrative methods of VR as necessary elements of their works when creating with VR to think about the existing relationship between display, viewing and being watched. This study took the VR art work "Inori" as the subject of case study. Through qualitative interviews, semi-structured interviews were conducted with the artists of this work, including Miwa Komatsu, Szu-ming Liu, general manager of HTC, Kay Huang, music creator, as well as with the audience. Based on VR 3I (Immersion, Interaction, Imagination) proposed by Burdea and Coiffet (2003) and related literature on VR text narrative methods, this study explores whether such six dimensions as immersion, interactivity, imagination, telepresence, region of interest and guiding the audience's attention are important elements in the artist's creation and influence the artist's performance and the audience's viewing experience. The results are as follows: Miwa Komatsu agrees that the six dimensions have a positive impact on VR art creation; Szu-ming Liu believes that the six dimensions have been taken into consideration in the creation, and they are specially designed for this purpose in the technical aspect; Kay Huang also agrees that VR has these characteristics and is specially designed for music creation; viewers believe that these six dimensions can make people more aware of the creator's creative mood and world than plane art or sculpture.

1 INTRODUCTION

With the rapid development of technology, the realization of virtual reality (VR) technology has opened up a broad space for artistic creation. The immersive space of VR brings a unique environment, which not only provides artists with new possibilities, new perspectives and creative methods, but also provides a world beyond the reach of graphic creation. VR has become a new form of creation, which can control and shape the works in the VR space, and re-create the spatial layout, so that users can have a strong sense of immersion and get a more diversified art appreciation experience.

This study takes the VR art work "Inori" as the subject of case study. Through qualitative interviews, semi-structured interviews were conducted with the artists of this work, including Miwa Komatsu, Szu-ming Liu, general manager of HTC, as well as with the audience. Based on VR 3I (Immersion, Interaction, Imagination) proposed by Burdea and Coiffet (2003), Bussell and Bilandzic's text narrative sense of existence and Oculus's region of interest and guiding the audience's attention and other VR text narrative methods, this study explores whether such six dimensions as immersion, interactivity, imagination, telepresence,

region of interest and guiding the audience's attention are important elements in the artist's creation and influence the artist's performance and the audience's viewing experience.

Therefore, the purpose of this study is twofold:

1. Will the immersion, interactivity and imagination of VR affect the artist's creative performance and the audience's experience of using it?
2. Will the existence sense of VR narrative methods, the region of interest and guiding the audience's attention affect the artist's creative performance and the audience's viewing experience?

2 VIRTUAL REALITY AND ITS APPLICATION

The creation method and process of digital art is far different from the traditional way, but the final requirement expressed by digital art, just like other arts, is the artistic sensory experience directly connected with the audience's mind. The rise of the new art category is understood as the reflection of human beings to contemporary society, the rise of digital art aesthetics and its inheritance of tradition and unique innovation. Starting from 2017, VR-based art creation began to rise rapidly around the world. Curated by Dajuin

---

*Corresponding Author

DOI 10.1201/9781003278474-14

Yao, China's first science-fiction-themed VR art exhibition "Mind Cosmos: Sci-Fi VR" appeared. At the same time, "Virtually Reality 3D Printed Artworks – Virtually Real", the first interactive art exhibition of virtuality and reality in the world, was jointly held by the Royal College of Art in the UK and HTC, marking the arrival of a new era of artistic creation.

In terms of personal creation, Ian Cheng presented Entropy Wrangler, a VR work at the London Art Fair in 2013, which explores automatically generated art in a live simulation approach. In 2017, Alejandro González Iñárritu won the first Academy Award for his VR film Flesh and Sand, which puts viewers into the roles of the participants to experience the lives of immigrants. In 2017, Laurie Anderson and Hsin-Chien Huang co-produced La Camera Insabbiata, which won the VR Best Experience Award of the newly established VR competition film project at the 74th Venice Film Festival. Ming-liang Tsai's VR film The Deserted, which collaborates with HTC, tries to change the framed screen world, using the traditional full-length shot to immerse the audience in the imaging and decide the order of viewing the image narrative. VR creation involves the audience and makes the audience's body an intermediary connecting the physical space and the virtual world. What VR artistic creation emphasizes is "how the audience imagines and thinks about the meaning of the artwork" (Chiu, 2019). With the rapid development of technology, the realization of VR technology has also opened up a broad space for artistic creation. It has become a new form of creation that can control and shape the work in VR space and recreate the layout of the space. The two-dimensional pattern of painting is extended to the three-dimensional space. Current 3D drawing is still in its early days and there is much room for improvement. The most advanced systems, such as Tilt Brush and Quill, only output grid geometry based on the artist's strokes, the system can't easily depict thick volume shapes, and tends to favor the collection of thin banded structures.

3   MIWA KOMATSU

3.1   Background of Miwa Komatsu

Miwa Komatsu (1984-) is a Japanese printmaker born in Nagano Prefecture, Japan. She specializes in drawing mythical creatures, the guardian deities of heaven and earth. Her works are themed with gods, mythical creatures and monsters who control death.

Miwa Komatsu likes the lines of copperplate etchings, so her early works are mostly copperplate etchings, which are mainly expressed in black or white, and mostly take life and death as the subject. But later, she wanted her work to be unique, not reproducible like print, so she destroyed the original copperplate etching (49 Days) that made her name as a copperplate artist. Later, she reexamined the relationship between gods and human beings, and the theme of her works shifted from life and death to prayer. As she traveled and met people of different cultures and faiths, she began to create colorful, vibrant paintings.

In 2014, her work Guardian Beasts of Heaven and Earth on the Chelsea Flower Show won the gold award and was collected by the British Museum. In recent years, Miwa Komatsu has used the creative mode of Live painting to make her work known to more people. For her, Live painting is an opportunity to convey ideas rather than show the process. The emphasis is on the meditative ritual that precedes painting. She believes that when the eyes are closed during meditation, the third eye opens, and meditation helps open the third eye to the guardian beast. The importance of prayer is conveyed in this process. Her works are full of religious and mysterious features.

3.2   Inori

In recent years, the global wave of digital aesthetics has been set off. Interactive art has changed the way of expression and appreciation of digital art, adding diversity and interest to art. Miwa Komatsu provided the concept of art and worked with Szu-Ming Liu to create INORI using the latest VR technology as an artistic extension. INORI describes the private creation process of Miwa Komatsu when her mind and body enter the universe of inspiration and step into the spiritual altar. INORI embodies the mental and body state of her painting process, so that the audience can feel the spiritual appearance of the creation. Ms. Kay Huang contributed to the production of the soundtrack. Miwa Komatsu believes that the eyes are the channel of energy used by the guardian beast, human and animal. The eyes can reveal a person's soul and power, and the mythical creatures can protect people and connect with people through the gaze of the eyes.

At the beginning, Miwa Komatsu meditates. With the guidance of Miwa Komatsu's voice, the audience can experience and follow Miwa Komatsu's creation process. The combination of motion and interaction makes it easier for the audience to understand Miwa Komatsu's inner world. "Prayer is a ritual common to many religions that conveys the deepest respect for God during meditation. Therefore, the act is not a performance, but a medium to convey ideas (Art Taiwan, 2018)". VR work experience makes up for the gap of "imagination" and "creativity" between ordinary people and artists. Through equipment and design, it helps the audience to enter another space more easily. Experiencers can get closer to the mind and vision of artists and understand their works.

4   VR NARRATIVE METHODS

The established film language seems to no longer work for VR and 360-degree films, leading to the development of a new narrative language. Filmmakers can no longer instruct the audience what to watch and the audience can decide what to watch themselves. Traditional editing is not applicable in VR, as the hop of time and space breaks the sense of immersion and presence during the scene switch (Michael Gödde, Dirk Siegmund, Frank Gabler & Andreas Braun, 2018).

Many scholars focus their attention on the connection between existence, narrative and interaction in VR. John Butcher (Bucher, 2017) stated that VR narrative is like any place in life, where we can walk to the place where the story happens and become the picture of the story. VR directors need to think about how to guide the audience rather than forcing the audience to watch the story from the perspective they decide. The content of the story is the most important and necessary, but the medium determines the way the story is presented, because different media have different narrative and presentation forms. The text of the story is abstract. Writers use literary techniques such as style, type, plot and narrative perspective to help readers read the story with their imagination, while stage directors and actors use body language or voice performance, combined with the design of props, sets, costumes and lighting to arouse the audience's emotional resonance to persuade the audience (Lin, 2019a, 2019b).

At present, the narrative language in VR has not been defined, but its pattern of manifestation can be found from the following four aspects. The first aspect is the sense of presence. Busselle and Bilandzic (2009) discussed the difference of the sense of presence caused by sensory stimulation involves the narrative itself. Roth (2017) believed that immersion is dependent on the objective criteria of software and hardware, which distinguishes immersion from presence. The sense of presence was later interpreted as a more subjective sense of physiology in an environment, influenced primarily by the content of the intervening world. Immersion can be considered as the quality of the medium, while the sense of presence is the user's experience, and deeper immersion may lead to a deeper sense of presence (Mirjam Vosmeer & Ben Schouten, 2017). Ryan (2001) proposed an interactive classification based on the dichotomy of ontology and exploration. Through exploratory interaction, the audience has no right to change the story, while through ontological interaction, the audience can change the story. The interaction is divided into three categories: exploring the virtual world and interacting with characters and manipulating objects. In traditional media such as books, television or movies, the way people receive story content is one-way. In VR, the audience can interact with the story and get some feedback. The third aspect is the regions of interest (ROI). The region of interest is the place that the director wants the audience to pay attention to. It is composed of symbols, roles and objects, including quantity, position and time of movement. The story is primarily delivered to the audience through ROI. The director needs to ensure that there are many stories around the audience, maintaining "spatial story density". Space becomes one of the elements that tell the story. The pace of the story in VR should not be too fast to ensure that the audience will not miss the story (Unseld, 2015). The distance of the ROI affects the depth of focus required by the eye.

In today's head-mounted displays, the adaptive requirements are fixed, which means that the audience can only adjust the focus requirements. In order to prevent eye strain, the ROI should be set at a comfortable distance to maintain the ROI for a long time (Oculus, 2016). The fourth aspect is to guide the audience's attention. VR directors cannot control the view of the audience, but they can use other elements to guide the audience. The first element is vision. Vision is easy to attract people's attention. The basic features of stimulating vision include line, color, the direction of motion and size and four secondary features include brightness start, brightness polarity, stereo depth, inclination, and lens movement. ROIs that use these features are more likely to grab the audience's attention. The second element is the primary character. The direction of the character's movements and the place of his gaze are both ways to guide the audience. The third element is the moving light and shading value. Fixed light is less effective. The fourth element is artificial visual cues. But this can easily affect the audience's sense of immersion. The fifth element is sound. Both the occurrence of events and the voice of character dialogue can attract the attention of the audience (Lin, 2019a, 2019b).

5 RESEARCH METHOD

Taking Miwa Komatsu's "Inori" as a case study, this study aims to explore the characteristics and narrative of VR artistic creation. Two artists, namely, Miwa Komatsu, Szu-ming Liu, the general manager of HTC VIVE, and an audience were interviewed to provide personal subjective opinions and feelings on the motivation and emotional factors of VR art creation. The main purpose is to understand the reason why artists choose VR as a creation tool, the way of expression of VR in creation, and the audience's perception of VR art after viewing the works. In this study, text analysis and in-depth interviews were used. The outline of the interview was refined from the literature content, and then the conclusion was summarized through the analysis of interview data.

There are three interviewees: Miwa Komatsu, Mr. Szu-ming Liu, and an audience who has watched "Inori". According to the interview outline, structured and semi-structured interviews were conducted on the 3I characteristics and narrative of VR.

In this study, data were collected by means of literature review, field participation and in-depth interview. VR 3I of Burdea and Coiffet, the sense of presence proposed by Busselle and Bilandzi and ROI and guiding the audience's attention mentioned by Oculus and other VR text narrative related literature were used as the measurement dimensions, including A1 immersion, A2 interactivity, A3 imagination, B1 sense of presence, B2 ROI and B3 guiding the audience's attention. The outline and question design are shown in Table 1.

Table 1. VR 3I, narrative dimensions, interview outline and questions.

Dimension No.	Interview outline	Questions
A1 immersion	P1. The creator (Miwa Komatsu)'s feeling about VR immersion	Q1. What's your opinion on VR immersion? Can it help you in your creation or inspire you to be more creative? Q2. When you choose VR, do you want the audience to be more immersed in your works?
	P2. Cognition of co-creator Szu-ming Liu (HTC)	Q1. As a co-creator, did you recommend VR because of its immersive experience? Q2. Do you think immersion is one of the most important elements of VR creation different from general creation? Q3. What kind of immersion experience do you want the audience to have?
	P3. The audience's feeling about immersion	Q1. Are you more immersed in Miwa Komatsu's "Inori" than in other types of artistic creation?
A2 interactivity	P1. Miwa Komatsu's idea of interactivity	Q1. Have you ever thought about the interaction between the painting and the audience when you were creating? Q2. What do you think can be conveyed to the audience by allowing the audience to interact with the paintings through VR?
	P2. The cocreator (HTC)'s cognition of interactivity in art creation	Q1. As a co-creator (HTC), do you think the interactivity of VR is an indispensable feature? Q2. What do you think is the biggest difference between VR creation with interactivity and graphic creation?
	P3. The audience's feeling about interactivity	Q1. Can the interactive experience when watching "Inori" better guide you to understand the work? Q2. Did interactivity increase or break your immersion during viewing?
A3 imagination	P1. Miwa Komatsu's cognition of imagination	Q1. Can VR creation enrich your imagination? Q2. Did VR change your way of creation or did you want to provide a different imagination space to the audience?
	P2. Co-creator Szu-ming Liu (HTC)'s cognition of imagination in artistic creation	Q1. When you created (HTC), did the imaginative nature of VR provide you with greater creative freedom or difficulty? Q2. What do you think is the biggest difference between "Inori" presented through VR and Miwa Komatsu's graphic works?
	P3. The audience's feeling about VR imagination	Q1. Did you increase your imagination after watching VR works? Did it give you different inspirations?
B1 sense of presence	P1. Miwa Komatsu's opinion on telepresence	Q1. Do you think it will help the audience to understand your work better if they participate in your creation from the first-person perspective or watch your work from the 360-degree space from the third-person perspective? Q2. What do you think about audience participation in your works?
	P2. Co-creator Szu-ming Liu (HTC)'s cognition of telepresence in artistic creation	Q1. As a co-creator, do you think increasing telepresence of the audience has a positive or negative impact on VR creation? Q2. What is your purpose in "Inori" for the audience to interact with the painting in the first-person perspective?
	P3. The audience's feeling about telepresence	Q1. Did you have deep telepresence when you watched the VR work?
B2 regions of interest	P1. The creator (Miwa Komatsu)'s feeling about ROI	Q1. Have you added any elements to the VR creation in order to make the audience better understand the information you want to convey? Q2. Is there any design in your works to increase the audience's interest in your works?
	P2. Co-creator Szu-ming Liu (HTC)'s cognition of ROI in artistic creation	Q1. How did you decide to extract some materials in order to increase the audience's cognition or interest in the work? Q2. Do you involve the audience in the interactive design to enhance their interest in watching or experiencing?
	P3. The audience's feeling about ROI	Q1. Are you particularly interested in any part when you watch VR creation?
B3 guiding the audience's attention	P1. The creator's feeling about guiding the audience's attention	Q1. Did you use sound elements to guide the audience to watch the work? Q2. In VR works, is the presentation of 360-degree space works designed to guide the audience to appreciate the information conveyed by the works?
	P2. Co-creator Szu-ming Liu (HTC)'s cognition of guiding the audience's attention in artistic creation	Q1. As a co-creator, do you think that guiding the audience to appreciate the work can enhance the audience's cognition of the work? Q2. What is the design method to guide the audience in "Inori"?
	P3. The audience's feeling about guiding the audience's attention	Q1. What do you think of the sound or lens that guides you in viewing? Are there other factors that can guide you to watch the VR work?

6 CONCLUSION

After summarizing the literature and analyzing the interview data, the research results are as follows:

7.1 Does the immersion, interactivity and imagination of VR affect the artist's creative performance, and the audience's use experience?

A1 immersion: Miwa Komatsu affirmed that the immersion and the way of creation of VR have a positive impact on thinking. Szu-ming Liu believes that VR can deepen viewers' understanding of his works, and he did not particularly work for the so-called immersion. However, because he was in charge of the HTC VR CONTENT, he was required to create fun for the audience when he was a producer, so there was a collision between the two perspectives. The audience believes that the immersion of VR enables them to truly enter the world created by the artist, and feel the mood of the artist in creation.

A2 interactivity: Miwa Komatsu expects her work to be interactive with the audience. Szu-ming Liu believes that interaction is important. Invisible interaction should be achieved, so that the invisible thing becomes visible, or the visible thing becomes invisible, so as not to interfere with the essence of the work. The audience believes that they can experience the process of the artist's creation through interaction.

A3 imagination: Miwa Komatsu believes that VR can really boost creative imagination and expects to continue to combine the latest technology such as VR to create works in the future. Szu-ming Liu said VR could improve the creative imagination and help ideas be shown in a way that everyone can understand. There's a lot of room for creativity, but it's difficult. This is a different way of creation, but the nature of creation remains the same. The audience believes that VR can promote imagination and help them show their ideas in an understandable way.

7.2 Does the narrative mode of VR, ROI and guiding the audience's attention affect artists' creative performance as well as the audience's viewing experience?

B1 telepresence: the creators agree that the telepresence can make the audience better understand the work, and the audience wearing the helmet can completely feel the world change in that space. The audience is deeply impressed by the telepresence. He believes that he can more clearly see the process of the artist's creation, his/her inner heart and what he/she wants to express to the audience.

B2 regions of interest: Miwa Komatsu said that the paintings in the VR work were specially painted, but not deliberately designed. She believes that everyone is free to decide which parts are they interested in. Szu-ming Liu also said that he didn't think much about the audience, but mainly focused on the world that the creator wanted to express. The audience also indicated that the ROI might vary from person to person.

B3 guiding the audience's attention: Miwa Komatsu said that the whole thing is well thought out, but not specifically designed. The work of guiding the audience's attention was left to the technicians. Szu-ming Liu believes that an interface to inspire the audience's behavior should be easy to understand, and the most natural behavior that triggers the audience is the most difficult. Intuition-inspired exploration is the most important subject in creation. The audience believes that guidance is very important, and he/she can quickly understand the artistic conception that the artist wants to express through the guidance of sound or camera conversion.

The results show that for artists, VR is a cross-border technology, and works should return to the essence. The characteristics of VR and the way of narrative can affect the creative performance of artists. The audience has a different feeling. Technology is used to complete the vision of the creator. VR is used to expand the vision of the audience, and it is used to allow the audience to experience the spiritual process of the artist, broaden their horizon and evoke resonance. The results of this study can be used as a reference for new media artists to create VR works in the future.

The interview was conducted during the outbreak of the epidemic, and the interview of music creator Kay Huang could not be carried out for the time being, so the conclusion was summarized based on the interview of two creators. Moreover, only a small number of audience participants could be found, so there are not enough samples. Therefore, researchers who want to conduct related topics in the future can increase audience samples to improve the reliability and validity of the study.

REFERENCES

Chiu, C. Y., "Ontological Event and (Syn) aesthetics in the Work of Art of Virtual Reality", Journal of Taipei Fine Arts Museum, (36), 2019.

Grigore C· Burdea, Philippe Coiffet, Virtual Reality Technology, 2nd Edition, 2003.

Lin, P. H., "True imitation, false reality – Sensory Experience Art Creation with Virtual Reality", Master's thesis, Department of Industrial Design, National Cheng Gung University, 2019a.

Lin, C. Y., "Investigating methods and principals for VR storytelling", Master's thesis, Master's Program in Digital Content & Technologies, National Chengchi University, 2019b.

Michael Gödde&Dirk Siegmund&Frank Gabler&Andreas Braun, CinematicNarrationinVR-RethinkingFilmConventionsfor360Degrees, Conference Paper · June, 2018.

Mirjam Vosmeer&Ben Schouten, A Project Orpheus A Research Study into 360° Cinematic VR, 2017.

Miwa Komatsu, Art Taiwan, Interview, Who is Miwa Komatsu?, H, 2018-07-26, /Interviewer: Art Taiwan,

Retrieved from https://arttaiwan.com/interview-09/, 2020. 12.25

Oculus, V. LLC, Oculus Best Practices, In: Technical Report., 2016.

Rick Busselle & Helena Bilandzic, Measuring Narrative Engagement, Media Psychology, Volume 12, Issue 4, 2009.

P.C.H. Roth, Experiencing Interactive Storytelling, PhD thesis, 2016.

Ryan, Marie-Laure, Narrative as virtual reality. Immersion and Interactivity in Literature, 2001.

Unseld, S. (2015). 5 Lessons Learned While Making Lost. Retrieved from https://www.oculus.com/story-studio/blog/ 5-lessons-learned-while-making-lost/2021. 04.20.

Wang, W., "Analysis of Composition", Journal of Jiaozuo University, 2005.

System Innovation in a Post-Pandemic World – Kin-Tak Lam et al. (Eds)
© 2022 Copyright the Author(s), ISBN: 978-1-032-24392-4

A novel deep neural network for air quality prediction

Yu-Shun Mao
Department of Electrical Engineering, National Sun Yat-sen University, Kaohsiung, Taiwan

Shie-Jue Lee*
Department of Electrical Engineering and Intelligent Electronic Commerce Research Center, National Sun Yat-sen University, Kaohsiung, Taiwan

ABSTRACT: Air quality prediction is considered a major issue for public health, and early warning and control can reduce the bad impact imposed by air pollution on the health of residents. This paper aims to predict air quality by a dual-path multichannel deep neural network model. The model employs convolution layers, gate recurrent unit, and attention mechanism to learn spatial features, temporal features, and other key features to make an effective prediction. The dual-path structure learns different dimensions of data, namely the temporal dimension and the feature dimension. This allows the structure to use temporal dependencies and feature associations to build high-level features. Also, time series decomposition is applied to obtain trend, seasonality, and residual components in the monitored time series data. The proposed model takes as input a collection of pollutant data, meteorological data, and decomposed components in a window of 168 consecutive hours. Various experiments' results show that our proposed system is superior to other systems in the accuracy of predicting air quality indices and pollutant concentrations.

1 INTRODUCTION

The World Health Organization reports that an estimated 23% of all deaths in 2012 (representing 12.6 million people) and 26% of deaths in children under age 5 were attributable to environmental risk factors, including pollution (Angela Bernhardt, Nathalie Gysi, 2016). Protecting cardiovascular and respiratory health can be partly ensured by a rapid phase-out of coal from the global energy mix. Many of the 2200 coal-fired plants currently proposed for construction globally will damage health unless replaced with cleaner energy alternatives. The phase-out of coal is proposed as part of an early and decisive policy package that targets air pollution from the transport, agriculture, and energy sectors, and aims to reduce the health burden of particulate matter (especially PM2.5) and short-lived climate pollutants, thus yielding immediate gains for society (Landrigan, P. J., et al. 2018).

In Taiwan, greater than 50% of the patients with lung cancer had never smoked. PM2.5 level changes can affect adenocarcinoma lung cancer incidence and patient survival (Tseng, C. H., et al. 2019). A study by Renzi et al. (2019) suggests a significant association of annual PM10 exposure with nonaccidental and cardiorespiratory mortality in the Latium region, even outside Rome and in suburban research shows that reducing PM2.5 concentrations and rural areas. The

study by Liu et al. (2017) suggests some positive associations between maternal exposure to ambient PM10 during the first two months of pregnancy and fetal cardiovascular malformations. In Schwartz, J. D., et al. (2018), the below the 2012 U.S. annual standard would substantially increase life expectancy in the Medicare population.

The purpose of this paper is to predict air quality for future hours using a dual path multichannel deep neural network. The remainder of this paper is organized as follows. Section 2 describes the proposed system, providing a prediction model framework and details. Section 3 presents experimental results. Finally, Section 4 concludes the paper.

2 PROPOSED METHOD

The main objective of this paper is to build an accurate air quality prediction system. The flow diagram of the proposed method is shown in Figure 1. Due to the complexity of the chemical factors, we use Pearson coefficient analysis to obtain the strongly correlated variable of air quality. Based on related works, we obtain four meteorological variables that have an impact on air pollutants. These are temperature, humidity, wind speed, and rainfall. We apply time series decomposition to low-level feature extraction. Time series decomposition is a method of time series analysis, which divided time series into trend,

*Corresponding Author

DOI 10.1201/9781003278474-15

seasonality, and residual. Time series decomposition provides a useful abstract model for thinking about time series generally and for better understanding problems during time series analysis and forecasting.

We proposed a dual path multichannel deep neural network which includes attention mechanism to build a dual feature filtering structure, and convolution layer and gated recurrent unit (GRU) layer for high-level feature extraction. We use two advantages of the structure to effectively increase the prediction performance of the model.

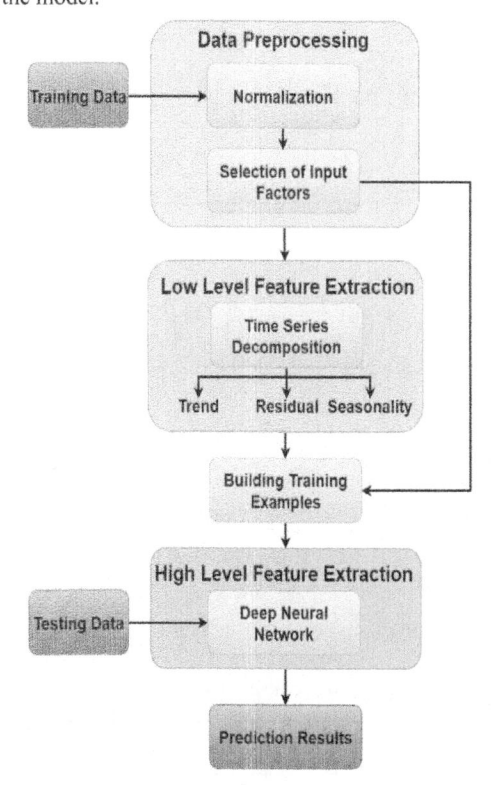

Figure 1. Flow diagram of the proposed method.

2.1 Deep neural network

After preprocessing, we obtain three low-level features, one chemical factor, and four meteorological factors. We design a dual-path multichannel deep network architecture that integrates a one-dimensional convolutional neural network (1D-CNN) (Kiranyaz, S., et al. 2021), GRU (Chung, J., et al. 2014), and Attention mechanism (Luong, M. T., et al. 2015).

(1) 1D-CNN

The forward processes of 1D-CNN layers are formulated as below:

$$C_i = \sigma((I \otimes k))_i$$

$$= \sigma \left(\sum_{m=1}^{M} \sum_{m=1}^{C} k_{m,c} I_{i+m,c} \right) \quad (1)$$

$$F_i = \text{Flatten}\,(C_i) \quad (2)$$
$$\hat{y} = (W * F + b) \quad (3)$$

where $\otimes$ denotes the convolutional operation, $\sigma(x)$ is the activation function, I is the input sequence, k is the kernel, k is the number of feature maps, i is the number of the kernel, and m is the number of input channels. We use the rectified linear unit (ReLU) as the activation function.

(2) GRU

The forward processes of GRU layers are formulated as below:

$$= \sigma\,(U_z x_t + V_z s_{t-1} + b_z) \quad (4)$$
$$r_t = (U_r x_t + V_r s_{t-1} + b_r) \quad (5)$$
$$h_t = \tan h\,(U_h x_t + V_h\,(s_{t-1} \odot r_t) + b_h) \quad (6)$$
$$s_t = z_t \odot h_t + (1 - z_t) \odot s_{t-1} \quad (7)$$

(3) Attention

The forward processes of the attention layer are formulated as below:

$$a_{ts} = \frac{\exp\left(\text{score}\left(\boldsymbol{h}_t, \bar{\boldsymbol{h}}_s\right)\right)}{\sum\limits_{SF}^{S} \exp\left(\text{score}\,(\boldsymbol{h}_t, \boldsymbol{h}_S F)\right)} \quad (8)$$

$$c_t = \sum_S a_{ts} \bar{\boldsymbol{h}}_S \quad (9)$$

$$a_t = \tan h\,([c_t; \boldsymbol{h}_t]) \quad (10)$$

$$\text{score}\left(\boldsymbol{h}_t, \bar{\boldsymbol{h}}_S\right) = \boldsymbol{h}_t^T \boldsymbol{W} \bar{\boldsymbol{h}}_S \quad (11)$$

Most of the papers on air quality prediction have proposed methods that focus on the establishment of time series dependence and ignore the effect of correlation between different input variables on the prediction performance. To solve this problem, we propose a two-path, multichannel deep network. The first path consists of three 1D convolutional layers, two GRU layers, and one attention layer. The first path of this network aims at learning spatial and temporal features by establishing short-term dependencies through 1D-CNN, and long-term dependencies through GRU, and then filtering mixed features through the attention layer to find the key features of this path. The second path is input to the network with the transposed data type. This path consists of three 1D convolutional layers and one attention layer. The purpose of the second path of this network is to learn association features by establishing interfactor dependencies in 1D-CNN, and then filter the association features by the attention layer to find the key features of this path. Finally, the key features of the two paths are filtered by the attention layer and input to the fully connected layer to obtain the network output.

3 EXPERIMENTS

In this section, we use the dataset collected at Zuoying by Environmental Protection Administration

(EPA) and Central Weather Bureau (CWB), Taiwan (https://data.epa.gov.tw/). The interval between two records is 1 hour, the time span is 2015/ 01/01 to 2019/12/31, the total number of records used is 43,824, and the number of variables is 18. We use the first 3-year data for training (2015/01/01 to 2017/12/31), next-year data for validation (2018/01/01 to 2018/12/31), and the last-year data for testing (2019/ 01/01 to 2019/12/31).

3.1 Experimental setup

Our proposed model is developed using Keras (https://keras.io/) and Tensorflow (https://www.ten-sorflow.org/) in Python. Scikit-learn (https://scikit-learn.org/) is used to build machine learning models. All experiments are deployed on a PC work station, and the work station configuration is Intel(R) Core(TM) i7-4770 CPU 3.40GHz with 16GB memory. The GPU, NVIDIA RTX 2070 8G, is used for acceleration.

We use root mean square error (RMSE), mean absolute error (MAE), correlation coefficient (Corr), and coefficient of determination (R^2), de fined as

$$RMSE = \sqrt{\frac{1}{N_{test}} \sum_{k=1}^{N_{test}} \left(Y_k - \hat{Y}_k\right)^2} \qquad (12)$$

$$MAE = \frac{1}{N_{test}} \sum_{k=1}^{N_{test}} \left|Y_k - \hat{Y}_k\right| \qquad (13)$$

$$Corr = \frac{\sum_{k=1}^{Nte} \left[(Y-Y)_k \times (\hat{Y} - \hat{Y})\right]_k^+}{\frac{1}{N_{test}} \sqrt{\sum_{k=1}^{N_{test}} (Y_k - Y_k)^2 \times \sum_{k=1}^{N\,test} \left(\hat{Y}_k - \hat{Y}_k\right)^2}} \qquad (14)$$

$$R^2 = \frac{\sum_{k=1}^{N_{test}} \left(Y_k - \hat{Y}_k\right)^2}{\sum_{k=1}^{N_{test}} \left(\hat{Y}_k - \tilde{Y}_k\right)^2} \qquad (15)$$

where N_{test} is the number of test examples, and $\hat{Y}_k, Y_k, \hat{Y}_k$, and Y_k are the predicted output, desired output, average of predicted outputs, and average of desired outputs, respectively.

3.2 Experimental results

The multistep ahead prediction performance of the Zuoying dataset is reported in Table 1 which gives RMSE, MAE, Corr, and R^2 comparison among support vector regression with RBF kernel (SVR-RBF), support vector regression with linear kernel (SVR-Linear), radial basis function (RBF) network, multilayer perceptron (MLP), and our proposed model for P2.5 prediction. As shown in these tables, our model is superior to other methods. The predicted values of PM2.5 at the future third hour by our method are shown in Figure 2.

Table 1. Prediction at future 1–3 hours for PM2.5 with the Zuoying dataset.

| Models | 1–3 h | | | |
	RMSE	MAE	Corr	R^2
SVR-RBF	7.062	5.475	0.747	0.759
SVR-Linear	6.932	5.319	0.740	0.765
RBF	7.411	5.706	0.725	0.732
MLP	6.942	5.191	0.742	0.763
Proposed method	**6.041**	**5.020**	**0.788**	**0.787**

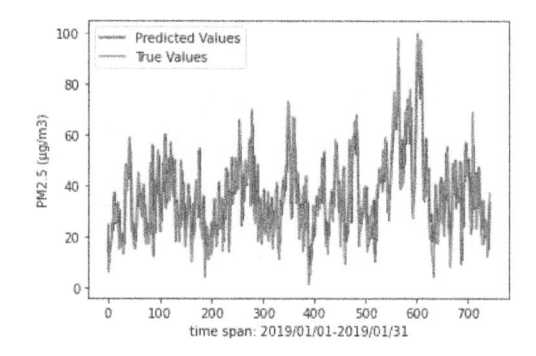

Figure 2. Predicted values of PM2.5 at Zuoying.

4 CONCLUSION

We have presented a dual-path multichannel deep neural network that can accurately predict the pollution concentration in the future. To prevent the deep neural network from over-fitting, relevance analysis is used to select highly correlated environmental factors for prediction. In the future, we would like to combine the Bayesian neural network and deep neural network. Training a Bayesian neural network via variational inference learns the distributions of parameters instead of the weights directly.

REFERENCES

Angela Bernhardt, Nathalie Gysi. (2016). World's worst pollution problems: the toxics beneath our feet. Zurich, 53p.

Chung, J., Gulcehre, C., Cho, K., & Bengio, Y. (2014). Empirical evaluation of gated recurrent neural networks on sequence modeling. arXiv preprint arXiv:1412.3555.

Landrigan, P. J., Fuller, R., Acosta, N. J., Adeyi, O., Arnold, R., Baldé, A. B., ...& Zhong, M. (2018). The Lancet Commission on pollution and health. The lancet, 391(10119), 462–512.

Liu, C. B., Hong, X. R., Shi, M., Chen, X. Q., Huang, H. J., Chen, J. H., ...& Sun, Q. H. (2017). Effects of Prenatal PM 10 Exposure on Fetal Cardiovascular Malformations in Fu-zhou, China: A Retrospective Case–Control Study. Environmental health perspectives, 125(5), 057001.

Keras, deep learning library, url: https://keras.io/.

Kiranyaz, S., Avci, O., Abdeljaber, O., Ince, T., Gabbouj, M., & Inman, D. J. (2021). 1D convolutional neural networks and applications: A survey. Mechanical Systems and Signal Processing, 151, 107398.

Luong, M. T., Pham, H., & Manning, C. D. (2015). Effective approaches to attention-based neural machine translation. arXiv preprint arXiv:1508.04025.

Renzi, M., Forastiere, F., Schwartz, J., Davoli, M., Michelozzi, P., & Stafoggia, M. (2019). Long-Term PM 10 Exposure and Cause-Specific Mortality in the Latium Region (Italy): A Difference-in-Differences Approach. Environmental health perspectives, 127(6), 067004.

Schwartz, J. D., Wang, Y., Kloog, I., Yitshak-Sade, M. A., Dominici, F., & Zanobetti, A. (2018). Estimating the effects of PM 2.5 on life expectancy using causal modeling methods. Environmental health perspectives, 126(12), 127002.

Scikit-learn, machine learning library in python, url: https://scikit-learn.org/.

Tensorflow, deep learning framework, url: https://www.tensorflow.org/.

The Environmental Protection Administration's environmental open data platform. url: https://data.epa.gov.tw/.

Tseng, C. H., Tsuang, B. J., Chiang, C. J., Ku, K. C., Tseng, J. S., Yang, T. Y., ...& Chang, G. C. (2019). The relationship between air pollution and lung cancer in nonsmokers in Taiwan. Journal of Thoracic Oncology, 14(5), 784–792.

System Innovation in a Post-Pandemic World – Kin-Tak Lam et al. (Eds)
© 2022 Copyright the Author(s), ISBN: 978-1-032-24392-4

Applying deep learning to the classification of exercise electrocardiography symptoms

Chun-Yen Chen
Department of Electrical Engineering, National Sun Yat-sen University, Kaohsiung, Taiwan

Shie-Jue Lee*
Department of Electrical Engineering and Intelligent Electronic Commerce Research Center, National Sun Yat-sen University, Kaohsiung, Taiwan

Hsiang-Chun Lee
Department of Internal Medicine, College of Medicine, Kaohsiung Medical University, Kaohsiung, Taiwan
Department of Internal Medicine, Kaohsiung Medical University Hospital, Kaohsiung, Taiwan

Ching-Yi Tsa, Su-Te Chen & Yu-Ju Li
Division of Cardiology, Department of Internal Medicine, Kaohsiung Medical University Hospital, Kaohsiung, Taiwan

ABSTRACT: A regular electrocardiogram (ECG) can show changes in the heartbeat but has the disadvantage of not being sensitive. By exercise, a dramatic increase in cardiac workload can be imposed, making the P- wave changes in the exercise ECG more sensitive than in a regular ECG and resulting in a high diagnostic sensitivity in the detection of atrial cardiomyopathy and paroxysmal atrial fibrillation. In this paper, we propose a deep learning system to learn whether the patient has suffered from atrial enlargement or atrial fibrillation. The system is divided into two main parts. The first part uses a convolutional recurrent neural network (CRNN) to identify the location of the P-waves in the patient's ECGs. The second part uses the P-waves found by CRNN to calculate the relevant parameters which are then used as input to a parallel bi-directional long short-term memory network (PBLSTM). The PBLSTM can analyze the P-wave parameters of different stages simultaneously and finally identify the patient's disease. Experimental results show that the developed system can identify atrial fibrillation more accurately than other systems.

1 INTRODUCTION

Prevention of atrial fibrillation (AF) is an important task to reduce the number of deaths caused by cardiovascular disease. Most people have atrial cardiomyopathy before the onset of AF, and the most common symptom is atrial enlargement of the right and left atria (Fragakis, N. et al. 2014, Pelliccia, A. et al. 2005). Atrial cardiomyopathy can be prevented from progressing to AF if it is effectively managed. However, because of the complexity and specificity of the test, echocardiography must be performed and interpreted by a specially trained technician or cardiologist and is a relatively expensive screening tool for large populations. Conventional electrocardiography (ECG) can demonstrate P-wave changes. which indicate the electrical remodeling of atria; the sensitivity is not good. Cost-effective diagnostic tools for high-risk groups have yet to be developed, and the current situation does not meet the clinical needs in this area.

Exercise can increase cardiac output by five times from the resting status, via multiple adaptations in the heart, including the increased venous return, increased heart rate, ventricular systolic force, stroke volume, and increased sympathetic nervous system activity. The acute increase over cardiac workload can be reflected by voltage changes of P- QRS-T waves in the ECG. For instance, changes of R wave amplitude in the ECG during exercise were first described by Simonson in 1953 (Simonson, E. et al. 1953). Later, changes in P-waves were described by Irisawa in 1966 (Irisawa, H. et al. 1966). The changes of P-waves during a submaximal exercise test include: (1) the peak P to Q interval shorting during exercise and lengthened in the recovery period; (2) the spatial magnitude increase during exercise; (3) a further augmentation of P-wave magnitude in the first minute of recovery.

A study demonstrated P-wave changes in exercise ECG of 20 cases with mitral stenosis, a group with

*Corresponding Author

DOI 10.1201/9781003278474-16

left atrial overload condition. The negative component of P-waves was increased from 16 to 19 of 20 in the post-exercise mitral stenosis group. In the normal control group, the negative component of P-wave did not increase after exercise (Yokota, M. et al. 1986). Another study compared P-wave duration and dispersion between two groups divided by the presence of paroxysmal AF (PAF). The mean left atrium diameter had no difference, and the maximum exercise P-wave duration and P-wave dispersion were greater than the rest measurements in the PAF group.

Therefore, P-wave changes during exercise can be of diagnostic value in the detection of atrial cardiomyopathy and in the prediction of PAF.

In this study, we apply the convolutional recurrent neural network (CRNN) model and the parallel bi-directional long short-term memory network (PBLSTM) model to analyze the P-wave changes of ECG during exercise. The developed system can be incorporated into clinical exercise ECG testing to identify the early-stage atrial myocardial disease and to help clinical control of the condition more effectively in order to prevent and reduce the likelihood of AF. The remainder of this paper is organized as follows. Section 2 describes the proposed system, providing a prediction model framework and details. Section 3 presents experimental results. Finally, Section 4 concludes the paper.

2 PROPOSED METHOD

The purpose of this study is to develop a diagnostic system for AF using artificial intelligence techniques. The project is divided into two major phases. The first phase lets our system capture the P-wave of the heartbeat from the exercise ECG about the atrial behavior of each patient at any stage. To achieve this goal, we need to understand how the ECG behaves in different exercise phases, and design a pre-processing program to remove noise and adjust the baseline for these situations, and then use artificial intelligence models to accurately capture the P-wave of each patient. In the second phase, the P-wave parameters of each patient are processed and calculated using the P-wave data obtained from each patient, and then the artificial intelligence model is used to diagnose whether the patient has atrial fibrillation or atrial enlargement, taking into account the dynamic changes in the P-wave parameters under different motion conditions.

2.1 Convolutional Recurrent Neural Networks (CRNN)

In this study, we used a CRNN to identify the location of P-waves. In this step, we take the normalized and baseline corrected ECG segments, and input them into our CRNN. Then use the P-wave labels given by KMUH to train the CRNN model that can capture the P-wave range of the ECG in the first step. The CRNN we use is a combination of a CNN and a BLSTM,

which is an advanced architecture of recurrent neural networks (RNNs).

We use CNN to fetch the features of different waveforms of each heartbeat. Because CNN can learn features by itself, many people use CNN for recognition that is, image recognition, voice recognition, and so on. CNN is constructed by convolution layer, pooling layer, and fully connected layer as shown in Figure 1. The convolution layer and pooling layer can fetch features of P-wave, and the fully connected layer finds out the P-wave location successfully. In each convolution layer, the following operations are performed:

$$C = R(E * k + b), \quad R(x) = \max(0, x)$$

where E is the waveform input of the ECG signal, k is a one-dimensional convolutional kernel of settable length, b is a constant value, is one of the activation functions called rectified linear units (ReLUs), and C is the feature map obtained after the convolutional operation of the ECG waveform and the convolutional kernel.

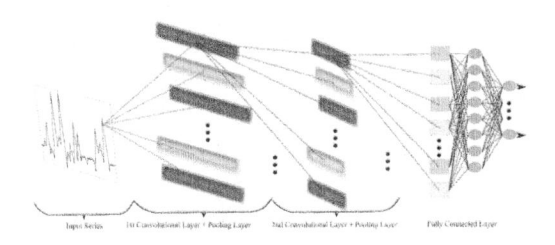

Figure 1. CNN architecture.

The LSTM model is an advanced architecture of RNNs. RNN has a storage unit that can efficiently store important features in a sequence, and RNN can input data of arbitrary length. The LSTM model can determine the input, output, and memory states by using three gate structures that adjust the signal flow, as shown in Figure 2. All three gates operate with the current input and the previous output, and control their output values between 0 and 1 by means of a sigmoid function.

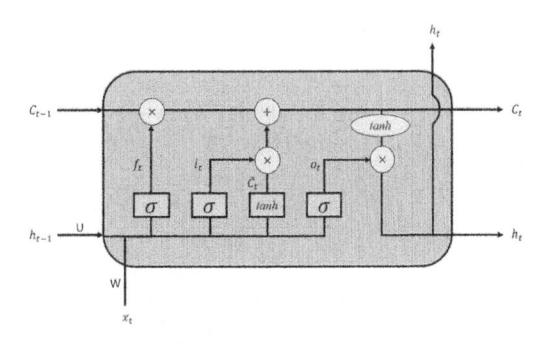

Figure 2. LSTM architecture.

The basic idea of bi-directional RNNs (BRNNs) is to propose two RNN for each training sequence, one forward and one backward, and both are connected to the same output layer. By combining the concept of bidirectional RNNs, LSTM allows the neural network to memorize not only the ECG sequence from backward to forward, but also from forward to backward (Yildirim, Ö . et al. 2018) so that it can capture the location of the P-wave segment better.

In addition, we designed a PBLSTM network architecture that can input data in parallel and remember the pre–post relationship. In our model, we use a masking layer to filter out null values. This architecture also adds an additional mechanism of Attention. The Attention mechanism will pay attention to the current input and the target, so that the model will focus on the part of the current input that has a higher correlation. In this study, it is the P-wave part of the ECG, and Attention can make our model pay more attention to the P-wave.

3 EXPERIMENTS

In this study, we collected the data of exercise ECG from 2017 to 2020 by using the research database of Kaohsiung Medical University Hospital (KMUH). From the variety of disease symptoms, we selected the data based on the three disease symptoms that were the focus of this study, namely normal, atrial enlargement, and AF. After ensuring that these data contained complete three-phase exercise ECG data, a total of 1308 patients were included.

3.1 Experimental setup

Our DNN model was developed using Keras (https://keras.io) and Tensor- Flow (https://www.tensorflow.org) whose libraries are written in Python. The experiments were carried out on a computer with Intel(R) Core(TM) i5-10400 CPU, 32GB RAM, and an NVIDIA GeForce RTX 2080 super graphics processing unit (GPU). We used the Stratified 5-Fold method for training and testing in our experiments. The stratified 5-fold method was used to conduct cross-validation.

Various evaluation criteria have been selected for the test data. These criteria include precision, recall, and F1-score. Note that macroaveraged criteria are used. Let TP, FP, TN, and FN indicate true positive, false positive, true negative, and false negative, respectively. The criteria for each category are defined as:

$$\text{Precision PRE}_i = \frac{\text{TP}_i}{\text{TP}_i + \text{FP}_i}$$

$$\text{Recall REC}_i = \frac{\text{TP}_i}{\text{TP}_i + \text{FN}_i}$$

$$\text{F1 score F1}_i = \frac{2 \times \text{PRE}_i \times \text{REC}_i}{\text{PRE}_i + \text{REC}_i}$$

The macro-averaged criteria are then defined as:

$$\text{Macro-averaged precision (Ma} - \text{P)} = \frac{\sum_{i=1}^{C} \text{PRE}_i}{C}$$

$$\text{Macro-averaged recall (Ma} - \text{R)} = \frac{\sum_{i=1}^{C} \text{REC}_i}{C}$$

$$\text{Macro-averaged F1 score (Ma} - \text{F1)} = \frac{\sum_{i=1}^{C} \text{F1}_i}{C}$$

where C denotes the number of categories.

3.2 Experimental results

In addition to experimenting with our proposed model, we also compare it with other proposed models to see whether our model can effectively extract the effective features for the diagnosis of symptoms. In these reference papers, most of the parameters and structures are constructed according to the original reference papers. Mei, Q. et al. (2020) use a CNN model with a five-layer convolutional layer using lead II waveform data input. He, R. et al. (2019) use a 12 lead waveform data input into two different architectures of multi-layer Residual block and bi-directional LSTM models; Ribeiro, A. H. et al. (2020) use a 12-lead waveform data input into a multi-layer residual block model with two inputs and two outputs. These three architectures have shown good results for the general ECG data presented in their paper. It can be seen that these architectures are effective in classifying general ECG data. The results of the comparisons with our proposed model are shown in Table 1.

Table 1. Comparison of macroaveraged scores.

	Precision (%)	Recall (%)	F1 (%)
Proposed study	**66.46**	**52.85**	**55.69**
Mei, Q. et al. (2020)	59.18	50.07	48.75
He, R. et al. (2019)	40.35	36.55	33.18
Ribeiro, A. H.et al. (2020)	26.02	33.33	29.23

4 CONCLUSION

We have presented a deep learning system for the Classification of Exercise Electrocardiography Symptoms. The patient's exercise ECG is fed into our system to decide whether the patient has symptoms of atrial enlargement or AF. The system uses CRNN to find the P-wave location of the patient's ECG. The P-wave found by CRNN is then used to calculate the relevant parameters and input them into the parallel deep learning architecture of PBLSTM. With our proposed deep learning system, we can effectively analyze the changes of patients in different phases simultaneously.

ACKNOWLEDGMENT

This work was supported by the grant MOST-108-2221-E-110-046-MY2, Ministry of Science and Technology, the NSYSU-KMU Joint Research Project (#NSYSUKMU 108-KN006 and #NSYSUKMU 110-KN002), and the "Intelligent Electronic Commerce Research Center" from the Featured Areas Research Center Program within the framework of the Higher Education Sprout Project by the Ministry of Education in Taiwan.

REFERENCES

Fragakis, N., Vicedomini, G., & Pappone, C. (2014). Endurance sport activity and risk of atrial fibrillation–epidemiology, proposed mechanisms and management. Arrhythmia & electrophysiology review, 3(1), 15.

He, R., Liu, Y., Wang, K., Zhao, N., Yuan, Y., Li, Q., & Zhang, H. (2019). Automatic cardiac arrhythmia classification using combination of deep residual network and bidirectional LSTM. IEEE Access, 7, 102119–102135.

Irisawa, H., & Seyama, I. (1966). The configuration of the P wave during mild exercise. American heart journal, 71(4), 467–472.

Mei, Q., & Liu, T. (2020, March). ECG diagnosis based on one-dimensional convolutional neural network. In IOP Conference Series: Materials Science and Engineering (Vol. 768, No. 7, p. 072110). IOP Publishing.

Pelliccia, A., Maron, B. J., Di Paolo, F. M., Biffi, A., Quattrini, F. M., Pisicchio, C., ... & Culasso, F. (2005). Prevalence and clinical significance of left atrial remodeling in competitive athletes. Journal of the American College of Cardiology, 46(4), 690–696.

Ribeiro, A. H., Ribeiro, M. H., Paixão, G. M., Oliveira, D. M., Gomes, P. R., Canazart, J. A., ... & Ribeiro, A. L. P. (2020). Automatic diagnosis of the 12-lead ECG using a deep neural network. Nature communications, 11(1), 1–9.

Simonson, E. (1953). Effect of moderate exercise on the electrocardiogram in healthy young and middle-aged men. Journal of applied physiology, 5(10), 584–588.

Yildirim, Ö. (2018). A novel wavelet sequence based on deep bidirectional LSTM network model for ECG signal classification. Computers in biology and medicine, 96, 189–202.

Yokota, M., Noda, S., Koide, M., Kawai, N., Yoshida, R., Mochizuki, K., ... & Sotobata, I. (1986). Analysis of the exercise-induced orthogonal P wave changes in normal subjects and patients with coronary artery disease. Japanese heart journal, 27(4), 443–460.

System Innovation in a Post-Pandemic World – Kin-Tak Lam et al. (Eds)
© 2022 Copyright the Author(s), ISBN: 978-1-032-24392-4

Exploring the key roles for digital content management

Chi Jung Lin*

Sanming University, Sanming, Fujian, China

ABSTRACT: With the development of technology, more and more intangible assets and intellectual property are produced and valued. The protection and management of these properties are also increasingly discussed. Early digital rights management focused on the design and discussion of a single role perspective. In this study, the three roles of the digital industry supply chain (platform provider, content producer, and consumer) are discussed together. Design and discussion of the life cycle of digital property and different pricing strategies for its competition model are presented. Finally, this research puts forward different opinions on and contributions to the implications of management and practice.

Keywords: Intangible assets, Digital rights Management (DRM), Intellectual property, Price discrimination

1 INTRODUCTION

With the development of smart devices such as the smart phone, personal digital assistant, mp3 player etc.; the behaviors of listening and reading have gradually changed. The content industry in culture and entertainment represents a significant business sector that produced about 7% of the U.S. GDP and exported around $80 billion worth of products in 1999 (Lang, Shang, & Vragov, 2009) . There are many digital rights management (DRM) technologies and methods that can be applied to various forms of digital content. However, the practicability and efficiency of these technologies and methods have been greatly challenged. DRM becomes more and more important as the digital contents are abundantly created.

The purpose of this paper is to provide an analysis about the pricing issue of digital content within DRM with duopoly model development (Choi, Bae, & Jun, 2010). We present an economic approach to analyze the circumstance under which pricing schemes may be optimal online serves. There are three key players in the digital content market: the content provider, the monopolistic online server/vendor, and the content consumer. How does the competition of two online servers affect the pricing strategy which the servers elect? In this research, we consider the search-based pricing and subscription pricing as two main factors in our duopoly model. We expect that the result will provide a deep view of pricing strategy of online servers in the dynamic competitive market.

2 DIGITAL CONTENT MANAGEMENT (DRM)

The DRM is a system that separately manages the digital content and the users' rights. The digital contents can be the files in texture, video, audio, picture, or other types. Before browsing those files, people must take the related permission in rights. There are two way to present the permission in rights, in file or hardware. Because the content data can be monitored by special decoder and debugger, and then the data can be grabbed. It is more effective to use a hardware-based solution rather than a software-based solution to make the rights data. The important data is stored in the temporary memory of the output hardware rather than in files, and then the grabbing is edificial if the stealer has no related hardware tools.

Although the DRM system is more powerful with hardware support, it is not popular to implement the hardware method. The reason is that the hardware solution needs the support from operating system, and that will be a conflict between the benefits of the companies. Therefore, the software-based method is still the most popular solution.

Traditional physical property management is based on its materials and physical characteristics. These management methods are not suitable for the management of intangible assets. Since the copying and transmission of digital content are very easy, today we have seen piracy that seriously violates copyright laws.

3 PRICE DISCRIMINATION

Using DRM to do differentiation is not a novel strategy. In traditional economics, there are some terms

---

*Corresponding Author

DOI 10.1201/9781003278474-17

covering doing things like that, for example, "price discrimination." Due to the mechanism of sales and promotion, it is difficult to propose a definition of price divergence that satisfies everyone.

First-degree price discrimination is perfect price discrimination – the producer succeeds in capturing the entire consumer surplus. This occurs, for instance, when consumers have unit demands exactly each consumer's reservation price and can prevent arbitrage between consumers. It then suffices for the consumer's reservation price.

Perfect price discrimination is unlikely in practice, either because of arbitrage or because of incomplete information about individual preferences. In the case of incomplete information about individual preferences, the producer may still be able to extract consumer surplus imperfectly by using the self-selecting devices. This is called second-degree price discrimination. Also the producer may observe some signal that is related to the consumer's preferences and use this signal to price-discriminate; this is termed third-degree price discrimination. The key difference between second-degree and third-degree price discrimination is that third-degree discrimination uses a direct signal about demand, whereas second-degree discrimination selects indirectly between consumers through their choice between packages. In our model, the problem of arbitrage can be ignored because of DRM.

4 ASSUMPTIONS AND ANALYSIS MODEL (DUOPOLY MODEL)

It is supposed that the market is oligopolistic and two major content suppliers share over than 80% market. The competition and pricing strategy that they choose are investigated. We assume that it is a market with complete information and the two suppliers are undifferentiated since they provide the same content to the consumers. Every supplier follows the decision structure below.

Stage 1: suppliers support one and only pricing scheme at a time. In case 1, two types of pricing schemes are restricted to frequency-based pricing and search-based pricing. In case 2, only subscription and search-based pricing are considered.

Stage 2: Each supplier decides on the specific prices, after observing the pricing strategy of opponent.

Stage 3: Customers make their own purchase decisions.

In first, this study considers the case when the supplier takes the frequency-based scheme then the case when the suppliers do not. We will use the concept of Nash-game perfectness to derive the equilibrium. By the way, the following Table 1 is shown all of notation which is used in the paper.

Table 1. Notation and the meaning.

Notation	Meaning
v	A valuation for a unit of content which takes a consumer time t to retrieve with mean μ and variance σ^2
t	The time which a consumer spends retrieving a unit content, is a normally distributed random variable
μ	The mean of t
σ^2	The variance of t
k	A risk parameter
y	The surplus of consumer which may be a random variable
c	The price which supplier pays to the content producer for each unit of time that the consumer uses the database
p_f	Subscription fee
p_c	Frequency-based fee
p_s	Search-based fee
$\pi(*)$	The profit function of the supplier
c'	The price which supplier pays to the content producer for each unit that the consumer buys from the database
q	Each consumer's demand of content
q_h	The demand of high-demand consumers
q_l	The demand of low-demand consumers
z	the proportion of consumers with elastic demand of total consumers
$1-z$	The remaining proportion of consumers
a	The willingness to pay for the first content
v'	The average valuation per consumer
n	The quantity of content

The Competition of Suppliers Considering Frequency-based Pricing

It is assumed that these valuations are distributed across the population according to a differentiable probability density function $f(v)$. The demand of consumers is elastic. Servers offer a frequency-based-based or search-based scheme to customers.

Proposition 1: If servers decide pricing strategy with specific prices; $c > 0$ and customers are different in terms of v. Then, we find that there are three pure strategy equilibrium of the game. Then we consider the market share A symmetric equilibrium is described that both servers choose a search-based policy and make zero profit. The remaining asymmetric two strategy equilibria are when one server offers a search-based pricing and makes positive profit, the other server offers frequency-based-based pricing and makes zero profits and vice versa

Proof:

In order to examine the sub-game perfect equilibrium of the game, we first consider the first stage of the game.

2. Stage 1:

We consider the first stage of the game. It could not be equilibrium for both firms to choose connect-based pricing policy, because one firm can benefit by deviating to search-based pricing. If one firm chooses frequency-based-based pricing and the other firm chooses search-based pricing, the firm which chooses frequency-based-based pricing cannot benefit by deviating because a deviation will only lead to zero profits. Similarly, if both firms are offering search-based pricing, no firm can strictly benefit by deviating to connect-based pricing. Again, switching to a frequency-based-based pricing will lead to zero profit for the deviating firm. Thus, the proposed equilibria are valid.

The proposition shows that an asymmetric equilibrium can appear with servers choosing different pricing schemes. Even when servers are undifferentiated, at least one server can get the positive profits. The proposition shows that consumers only differed in term of v

1. Stage 2:

If both servers offer the same pricing schemes, they will make zero profits. Therefore, consider the case when the servers offer different pricing policies. A customer will buy the product from a firm offering searchbased pricing if:

$$v - p_s > v - \mu p_c - \sigma^2 k p_c^2 / 2 \qquad (1)$$

$$p_s < \mu p_c + \sigma^2 k p_c^2 / 2 \qquad (2)$$

Since this term is independent of v all consumers will buy from only one firm and therefore only one firm will make positive profits. We note that the firm offering frequency-based-based pricing will charge $p_c^* \geq c$. So, if one server offers a frequency-based-based pricing, another server using a search-based price

$$p_s^* = \mu c + \frac{\sigma^2 k c^2}{2} - \varepsilon \qquad (3)$$

We can charge a price where ε is small. The profits in this case are

$$\prod_s^* \geq \prod_s = \left(\frac{\sigma^2 k c^2}{2} - \varepsilon\right) \int_{\mu c + \sigma^2 k c^2/2}^{\infty} \\ \left(v - c\mu - \frac{k\sigma^2 c^2}{2}\right) f(v) d > 0 \qquad (4)$$

Thus, the firm offering the search-based-pricing will make positive profits. Table 2 is the payoff matrix.

Table 2. Payoff matrix.

	frequency-based-based pricing	search-based pricing
frequency-based-based pricing	π_c, π_c	$0, \pi_s$
search-based pricing	$\pi_s, 0$	$0, 0$

Table 3. Market share matrix.

	frequency-based-based pricing	search-based pricing
frequency-based-based pricing	M_c, M_c	$0, M_s$
search-based pricing	$M_s, 0$	M_c, M_c

The Competition of Servers Not Considering Frequency-based Pricing

It is assumed that there are two segments of consumers in the market, z-type consumers with elastic demand and (1-z) consumers, the remaining consumers in the market. Servers offer subscription or search-based scheme to consumers.

Proposition 2 If servers decide pricing strategy with the specific prices, $c > 0$. If we have $a < v < \frac{a^2}{2}$ and $a > 2$, there is only one pure strategy equilibrium game. One server offering a search-based pricing makes positive profits and the other server offering subscription pricing also makes positive profits.

Proof:

In order to examine the sub-game perfect equilibrium of the game, we first consider the second stage of the game.

1. Stage 1:

If both servers offer the same pricing schemes they make zero profits. Therefore, consider the case when the servers offer subscription and search-based scheme, it is profitable to offer search-based scheme to (1-z)-type consumers and offer subscription scheme to z-type consumers when $a < v' < \frac{a^2}{2}$ and $a > 2$.

Consumers with elastic demand will buy from a server offering subscription in the fee is required in the interval: $c' < p_f^* \leq \frac{a^2}{2}$. The consumers with inelastic demand will buy the product of the server offering search-based pricing with $c < p_s^* \leq v'$.

When one server offering the subscription pricing sets the price as $p_f = \frac{a^2}{2}$, it earns a profit of $\pi_f = \frac{a^2 z}{2} - \left(a - \frac{a^2}{2}\right) zc'$. The other sever offering the search-based-pricing sets the price as $p_s = v'$, it ears a profit as following:

$$\pi_f = (v - c')(1 - z)/2 \qquad (5)$$

Thus, under this circumstance, if servers offer different pricing scheme and set the price, both of them can make positive profit. Table 4 is the payoff matrix.

Table 4. Payoff matrix.

	Subscription pricing	search-based pricing
Subscription pricing	$0, 0$	π_f, π_s
search-based pricing	π_s, π_f	$0, 0$

2. Stage 2:

We consider the second stage of the game. We don't have the equilibrium for both firms to choose the same pricing policy. They will get into Bertrand competition and one server can benefit by deviating to another pricing. If one server chooses search-based pricing and the other chooses subscription pricing, no firm can benefit by deviating. The deviation will only lead to zero profit.

Therefore, we consider if it is possible for one or both servers to set price to attract some consumers form each segment? We define that the profit from setting a price equal to the willing-to-pay of the z-type consumers provided a higher profit to the server offering a subscription. Similarly, the server offering search-based never want the profit to be less than the profit from setting a price equal to the willing-to-pay of z-type consumers. In order to obtain these profit lower-bounds, there must be no deadweight loss. If the server offering the subscription wants to attract any $(1-z)$-type consumer, there must be some dead-weight. All $(1-z)$ consumers must buy the product of the server offering the search-based scheme. The server offering the subscription to achieve its profit lower bound it must sell to all z-type consumers. Thus, the proposed equilibrium is valid.

5 EMPIRICAL ANALYSIS

We analysed the pricing methods of three categories of digital content: on-line music, on-line game and e-learning. The result is as follows

5.1 Pricing of on-line music in America

There are two popular pricing modes of on-line music nowadays, one is paying for single song the other is a subscription for a month or a year. The mainstream is pay per song, about 65% in the market. Taking the leading server, iTunes, for example, it only offers the first scheme. However there are more and more consumers and competition, the demand of subscription is appearing. This trend makes some servers provide the subscription to satisfy consumers with dissimilar need.

After the success of iTunes, many companies invest in the growing market, including AOL's MusicNet, RealNetworks' RealRhapsody, Wal-Mart, and Sony Connect. Microsoft also opened an online music shop on MSN in September, 2004.

When it comes to their pricing strategy, limited to the cost of copyrights royalty to content owners, the price of on-line music is between 0.79 to 0.99 U.S. dollars. The pricing methods are in Table 5. Regarding the number of songs, because many record companies want to cooperate with these servers, there are more than five hundred thousand songs in their databases.

Table 5. Pricing method of on-line music.

Server	Company	Pricing Method
iTunes	Apple	pay per song
RealRhapsody	RealNetworks	pay per song and monthly subscription
Musicmatch	Musicmatch	pay per song and monthly subscription
Napster 2.0	Roxio	pay per song and monthly subscription
MusicNet (new version)	AOL	pay per song and monthly subscription
Wal-Mart Connect	Wal-Mart	pay per song
	Sony	pay per song
MSN Music	Microsoft	MSN Music Microsoft pay per song

5.2 Pricing of on-line game in Taiwan

The largest five on-line game companies in Taiwan are Soft-World, SoftStar, Wayi, Gammnia and InterServe and they capture almost 80% consumers in the market. The payment methods of them are based on the "point", a point is about one N.T. dollar. Consumers can buy point-cards in convenience stores or internet cafes or by credit cards.

On-line game companies offer two choices for consumers, paying for hours or monthly subscription. We find the payment method is via "points", but the basic ideas are two strategies: frequency-based-based and subscription. The pricing methods are shown in Table 6

Table 6. Pricing method of on-line game.

Name of the Game	Company	Pricing Method
Ragnarok Online(RO)	Soft-World	Frequency-based-based payment and monthly subscription
gamania lineage	NcSoft	Frequency-based-based payment and monthly subscription
Seal Online	Taiwan Index Corporation	monthly subscription
Dragonraja	TWP	Frequency-based-based payment and monthly subscription
Stoneage	Wayi	Frequency-based-based payment and monthly subscription
Chain of Life	Soft-Star	Frequency-based-based payment and monthly subscription

5.3 Pricing of E-learning in Taiwan

The major types of e-learning content in Taiwan are English and integrated tuition. The integrated tuition includes before-school-age education, junior-high-school teaching materials and advance to a higher school education.

The popular servers in the market and their pricing methods are in Table 7 the paying ways of e-learning are time-based payment and course-based payment. The course-based payment means that there is different price for different course and consumers have to pay for each course. This kind of payment is just like the search-based pricing mentioned before.

Table 7.   Pricing method of E-learning.

Server	Content Type	Company	Pricing Method
kid.com.tw	English and integrated tuition	Strawberry Software	subscription
Kid Castle www.uc520. com	English integrated tuition	Kid Castle InterServe	subscription course-based payment and subscription
TKB e-learning center	advance to a higher school education	Taiwan Knowledge Bank	course-based payment

While a server pays royalty, based on how many content consumers acquire, to the content owner, it would be less flexible. Frequency-based-based is one of pricing choices for the first kind of server but is unsuitable for the second kind of server. Taking on-line game "Chain of Life" for instance, it is made by Soft-Star and the copyright also belongs to it, so Soft-Star offers Frequency-based-based payment and subscription to consumers. When surveying these on-line game servers, we can find most servers offer two different pricing schemes to consumers, and this situation responds to Proposition 1.

For on-line music servers, because of their cost structure, they offer pay per song or subscription schemes to consumers. Most servers offer both these two choices, this situation responds to Proposition 2.

6   CONCLUSION AND DISCUSSION

From the survey of digital content servers, we find that different server with different cost structure has dissimilar choices of pricing strategy. So we present two cases, the server taking the frequency-based pricing into consideration and the other does not. Different market situations, Duopoly market is discussed in this study. In duopoly market, no matter in any case, it is profitable if two servers use different pricing to each other. The different payment method would attract different kind of people, who may have different quantity of demand or different type of demand.

There is a large number of multimedia and visual content on the Internet, which can be used for many applications, such as education, entertainment, academia, and research. The copyright and intellectual property rights of these multimedia contents need to be preserved.

To solve the data management problems of digital twins in the product life cycle, including data storage, data access, data sharing, data authenticity, and data coverage (virtual products). A peer-to-peer network is established to connect every participant in the product life cycle. The transaction records all the actions of the product digital twin among the participants. Sensor data between physical products and virtual products are also recorded through transactions. All transactions are stored in blocks linked by cryptography to form a blockchain. Time stamping involves the entire process of marking the time of occurrence. The case study is associated with the digital twin of the turbine to demonstrate the effectiveness of the proposed data management method. The results show that the proposed method can solve the above data management problems at the same time, that is, data can be stored in blocks, the accessed data needs to be verified, data sharing through peer-to-peer networks is effective, and the traceability of data coverage can be avoided. The authenticity can be guaranteed.

However, the documents stored in the blockchain will reduce the query efficiency of the blockchain, which is not conducive to the data management of the digital twin and will be studied in depth in future work.

Besides, this article only solves the data management problem of the product digital twin. In future work, we can pay more attention to the digital twin of the floor workshop, manufacturing system, and supply chain. (Huang, Wang, Yan, & Fang, 2020)

Digital rights management is an important topic in the network environment. A new paradigm of digital rights management is based on blockchain, which supports correct digital rights protection content and provides services to the correct users in the correct way. In the future, we will strengthen support based on the ether The work of digital rights management and trading coins to support a new and promising vision: the right content, the right way, and the right value to serve the right users. (Ma, Jiang, Gao, & Wang, 2018).

REFERENCES

Choi, P., Bae, S. H., & Jun, J., 2010. "Digital piracy and firms' strategic interactions: The effects of public copy protection and DRM similarity". Information Economics and Policy, 22(4), 354–364.

Huang, S., Wang, G., Yan, Y., & Fang, X., 2020. "Blockchain-based data management for digital twin of product". Journal of Manufacturing Systems, 54, 361–371. doi:https://doi.org/10.1016/j.jmsy.2020.01.009

Jean Tirole. "The Theory of Industrial Organization," MIT Press Books, The MIT Press, edition 1, volume 1, number 0262200716, February 1988.

Lang, K. R., Shang, R. D., & Vragov, R.,2009, "Designing markets for co-production of digital culture goods". Decision Support Systems, 48(1), 33–45.

Ma, Z., Jiang, M., Gao, H., & Wang, Z.,2018. "Blockchain for digital rights management." Futur. Generation Computer Systems, 89, 746–764. doi:https://doi.org/10.1016/j.future.2018.07.029

Pigou, A. C., 1924. The economics of welfare: Transaction Publishers.

System Innovation in a Post-Pandemic World – Kin-Tak Lam et al. (Eds)
© 2022 Copyright the Author(s), ISBN: 978-1-032-24392-4

Applying fuzzy Delphi method to explore and formulate sustainable development indicators for school's green campuses in Taiwan

Con-Rong Wang
College of Fine Arts and Design, Jimei University, Xiamen, Fujian, China

Vivien Yi-Chun Chen*
Department of Architecture, Fujian University of Technology, Fuzhou, Fujian, China
Feng Chia University, Taichung City, Taiwan

Artde Donald Kin-Tak Lam
Fujian University of Technology, Fuzhou, Fujian, China

Jung-Chen Huang
Department of Environmental Engineering, National Cheng Kung University, Tainan City, Taiwan

ABSTRACT: Sustainable development is a global trend, prompting the development of green school buildings. Green building and low-carbon campus promotion in junior high schools and elementary schools were researched in this study to facilitate improved environmental protection, energy-saving, and carbon reduction as well as the construction of green, low-carbon campuses. Questionnaires were administered to experts from different fields regarding the decision sequence for constructing a green, low-carbon campus in junior high schools and elementary schools. The results provide a reference for architectural planning and implementation for schools and for researchers. The government can use them to inform appropriate policies and to establish priorities and indicator weights for promoting green, low-carbon school campuses. Educational resources can thus be reasonably allocated, green building quality on campuses enhanced, and energy-saving and carbon-saving measures implemented.

Keywords: Green Campus; Sustainability; Green School Building; Fuzzy Delphi Method

1 INTRODUCTION

In accordance with the global sustainable development trend, in 2002, Taiwan planned the Challenge 2008—National Development Plan, and the Ministry of Education planned the Sustainable Campus Project with the goal of creating green, sustainable campuses. Based on the favorable results obtained from the above projects on green building, energy-saving, water saving, and ecology and environmental protection, it expanded its practice and launched the Eco-city Green Building Promotion Project in 2008 to incorporate sustainable cities into Taiwan's green building projects. In 2010, Taiwan incorporated smart green buildings into four major emerging smart industries (i.e., cloud computing, smart electric cars, smart green building, and industrial inventions and patents) and into the Eco-city Green Building Promotion Project. In December 2010, the Smart Green Building Promotion Project was passed, according to which, smart green

campuses were to be promoted each year, and the regulation was amended such that new public constructions with a project cost of more than NT$50 million would be required to apply for a a smart building mark. In 2016, the government promoted the Sustainable Smart City—Smart Green Building and Community Promotion Project, striving for increased energy-saving and enhanced indoor environmental quality as well as reduced environmental impact.

2 EXPERIMENTAL SECTION

Regarding sustainable campuses and green school building planning, (C.M. Tang, 2014; C. Ghent et al., 2014) stated that sustainable school planning and design must consider the following factors: energy efficiency, environmental impact, resource protection, and air quality. (H. Meiboudi et al., 2014; C.J. Kibert, 2012; J .C. Wang, 2016). Shih et al. (2014) contended that sustainable schools should reduce waste; use materials and resources that can be repaired, renewed, and reused; establish renewable energy sources; utilize

*Corresponding Author

environmental climate features to reduce reli ance on energy; construct a healthy and safe learning environment; and connect school education with community resources (AASHE, 2019). Yeh et al. (2017) analyzed the green building planning for junior high schools and elementary schools regarding the building site and eco-environment, and found that school sites, compared with those of general buildings, have larger areas, richer ecology, more greenery, and more permea ble layers. They also concluded that green buildings are an alternative for the urban areas that lack green lands. During the early stages of the planning, the campus life-cycle assessment plan should be considered comprehensively to avoid huge maintenance and repair fees in the future (C.M. Tang 2014, AASHE 2019). Resources such as flexible space, maintenance and management, durability design, large-area campus sports fields, and idle space generated by social changes must be exploited (C.M. Tang 2014, T. Savelyeva et al. 2012, X. Luo et al. 2020).

The education reform movement has led to many forms of open concepts in the operation of junior high schools and elementary schools. Tang (2014) outlined the following five aspects to be considered in planning sustainable campuses with green buildings.

1) Emphasizing the relationship between school and community, and seeking reasonable transparency and openness in opinion exchange and participatory ex pression.
2) Emphasizing the cycle of management of resources and energy, resource recycling and reuse, rainwater regeneration and reuse, the ap plication of renewable energy, energy-saving design measures, and water-saving and electricity-saving utilities.
3) Emphasizing the sustainable management of the construction foundation, including the improvement of the surface soil, multilayer greenery, and planning ponds for scenery or teaching.
4) Val uing eco-environments, including establishing a field, for education purposes, where fallen leaves and food waste are used as compost or where symbi otic animals are kept.
5) Valuing healthy building en vironments, including adaptations for the local natu ral climate and the use of eco-friendly construction materials (X. Luo et al. 2012, J.C. Wang et al. 2019).

The design of campus space is no longer restricted by fences or rigid concrete buildings. Campuses can be more localized, diverse, green, and esthetic, with a garden-like atmosphere, and humanized. Edu- cation methods can be rendered more flexible, more open, and more relevant to daily life. Green school campuses are characterized by resource saving, substance recycling, low pollution, and low noise levels (H. Yan et al. 2009). School administration and teachers and students promote environmentalism on- and off-campus, for example, through teaching that focuses on environmentalism, participation in com- munity cleaning and greening, civil and governmen tal activities, environmental teaching, expansion of local and outdoor education, establishment of trails for environmental education in communities and schools, and activities such as immersive teaching and green curriculum development (C.M. Tang 2014).

3 RESULTS AND DISCUSSION

This study performed a literature review and employed the Fuzzy Delphi method to obtain research results.

To discuss factors involved in promoting green, low-carbon campuses in junior high schools and elementary schools in Taiwan, in the literature review section, literature on low-carbon schools, green school buildings and assessment systems, sustainable campuses, and green schools in Taiwan and abroad was collected as the basis for research. The assessment factors and framework selection were based on the principles of objectivity, scientificalness, predictability, and comparableness (J.B.M.B. Sanfo, 2020). Relevant literature on green, low-carbon campuses was examined. We preliminarily organized 56 assessment factors for promoting green, low-carbon campuses (Table 1) into the following six major categories:

1. sustainable site planning;
2. campus maintenance and lifespan management;
3. life and learning;
4. ecological environment;
5. energy and carbon-saving design, planning, and application;
6. healthy campus environment.

3.1 Preliminary test for the reliability of the Fuzzy Delphi expert questionnaire

Seven experts and scholars participated in the preliminary test of the questionnaire for experts. This study adopted internal consistency Cronbach's α values to measure reliability. Cronbach's α value of each assessment factor should be no less than 0.7. Reliability analysis revealed that overall Cronbach's α value was 0.845 (Table 1). Cronbach's α of standardized items was 0.853. Cronbach's α value of each item was between 0.836 and 0.855 (Table 2), indicating favorable internal consistency and reliability. The questionnaire exhibited stability, so a formal questionnaire based on the tested version could be implemented.

3.2 Selecting a threshold value screening evaluation factor

Threshold values affect the screening numbers of assessment factors. Generally, threshold values are

Table 1. Fuzzy Delphi expert questionnaire preliminary test reliability statistics.

Cronbach's α	Cronbach's α based on standardized iitems	No. of items
0.845	0.849	56

Table 2. Fuzzy Delphi questionnaire for collecting the expert's statistics.

Assessment Factor	Fuzzy Delphi expert questionnaire reliability preliminary test reliability statistics.				Fuzzy Delphi questionnaire for expert's reliability preliminary test overall statistics									
	Scale Mean After Removal of Items	Scale Scale Standard Deviation After Removal of Items	Modified Total Item Correlation	Cronbach's Alpha After Removal of Items	Most Conservative Cognition Value	Most Optimistic Cognition Value			Geometric Mean		Assessment Value		Expert Consensus Value
					Min (CLi)	Max (CUi)	Min (OLi)	Max (OUi)	CMi	OMi	Single value	MiZi	Gi
School location selection	442.86	503.143	0.648	0.838	3	6	5	9	5.05	7.65	6.65	1.60	5.74
Establishment of overall demand plan and environmental assessment	442.71	533.238	0.201	0.846	4	7	7	9	6.02	8.66	7.49	2.64	7.34
Campus energy-saving management policy	444.71	519.571	0.463	0.842	4	8	8	10	6.15	8.81	7.28	2.66	7.48
Lifespan extension plan for facility maintenance management	444.43	527.952	0.234	0.845	3	8	7	10	5.82	8.80	7.47	1.98	7.45
Life-cycle assessment of the buildings on campus	445.00	557.000	−0.302	0.855	3	8	6	10	5.14	7.91	6.54	0.77	6.80
Building economy of scale	443.14	539.476	0.000	0.850	2	6	5	10	5.25	7.02	6.54	0.77	5.73
Environmentally friendly school building planning	443.71	511.905	0.480	0.841	5	8	8	10	6.04	9.03	7.37	2.99	7.54
Create creative campus features	444.29	548.238	−0.188	0.851	4	8	7	10	6.10	8.80	7.30	1.70	7.49
Digital environment and learning	443.29	522.905	0.302	0.844	2	8	6	9	5.03	7.05	8.02	0.02	6.52

(*continued.*)

Table 2. Continued

Assessment Factor	Fuzzy Delphi expert questionnaire reliability preliminary test reliability statistics.				Fuzzy Delphi questionnaire for expert's reliability preliminary test overall statistics								
	Scale Mean After Removal of Items	Scale Scale Standard Deviation After Removal of Items	Modified Total Item Correlation	Cronbach's Alpha After Removal of Items	Most Conservative Cognition Value	Most Optimistic Cognition Value			Geometric Mean		Assessment Value		Expert Consensus Value
					Min (CLi)	Max (CUi)	Min (OLi)	Max (OUi)	CMi	OMi	Single value	MiZi	Gi
Participation of teachers, students, and the public	443.29	540.571	−0.011	0.849	3	8	6	10	5.03	7.81	6.59	0.78	6.76
Overall use of green construction principles in the school and the community	443.71	496.571	0.684	0.836	3	8	6	10	5.53	8.19	6.88	0.66	6.94
Green procurement	443.71	511.905	0.804	0.839	3	7	6	10	5.07	8.11	6.75	2.04	6.52
Green transport	444.14	515.810	0.363	0.843	3	7	5	9	4.54	7.52	6.16	0.98	6.01
Media on campus environmental protection	443.00	540.667	−0.008	0.849	3	6	4	8	4.76	6.58	5.57	−0.18	5.35
Green low-carbon courses and education	443.14	547.476	−0.0183	0.850	3	8	6	10	5.66	8.57	7.24	0.91	7.05
Maintenance of favorable campus ecology and restriction of development	443.86	545.143	−0.130	0.850	4	8	7	10	5.89	8.66	7.34	1.77	7.44
Repair of ecoenvironmental damage 9	443.57	519.952	0.455	0.842	3	8	7	10	5.07	7.90	6.30	1.83	7.23
Land restoration	443.14	503.810	0.694	0.838	2	8	6	10	4.89	7.68	6.42	0.79	6.70
Assessment of the impact of development on local ecology	443.29	506.238	0.497	0.840	4	8	7	10	5.60	8.37	6.89	1.77	7.36

Biodiversity	442.71	543.238	−0.064	0.850	4	8	7	10	5.92	8.80	7.18	1.88	7.46
Indigenous flora and plant diversity	444.14	536.810	0.097	0.847	4	8	7	10	6.07	8.59	7.63	1.52	7.45
Organic biofarming	444.29	502.238	0.687	0.837	4	7	7	9	4.92	7.71	6.27	2.79	7.00
Light hazard prevention	444.29	536.571	0.029	0.851	2	7	5	10	4.50	7.46	6.47	0.96	5.99
Biopond establishment	443.86	534.476	0.174	0.846	2	7	6	10	4.28	7.49	5.86	2.21	6.35
Ecological multi-layer three-dimensional greening	443.43	520.286	0.460	0.842	4	8	7	10	5.68	8.49	7.10	1.81	7.39
Extent of campus greening	444.00	522.000	0.446	0.843	4	8	7	10	5.96	8.89	7.36	1.93	7.48
Old tree protection	443.29	525.571	0.294	0.845	4	8	7	10	5.91	8.78	7.22	1.87	7.46
Reduce building coverage and underground excavation	443.71	525.238	0.281	0.845	4	8	6	10	5.54	8.05	6.75	0.51	6.91
Facilities for flood prevention, water retention, and water storage	442.86	505.810	0.599	0.839	4	8	6	10	5.34	8.24	6.89	0.90	6.91
Water permeable area	443.43	529.619	0.201	0.846	4	8	7	10	5.68	8.51	7.38	1.83	7.39
Soil conservation	443.00	538.000	0.083	0.847	3	7	6	10	5.09	7.86	6.47	1.77	6.49
Energy-saving building shell design	444.57	502.286	0.739	0.837	3	8	6	10	5.87	9.04	7.56	1.17	7.18
Air conditioning and lighting energy saving	444.43	523.286	0.192	0.847	3	7	5	10	5.19	7.98	6.59	0.79	6.24
Energy-saving lighting equipment and system	444.14	553.143	−0.245	0.853	4	8	7	10	5.80	8.81	7.10	2.01	7.45
Manageable green renewable energy	443.71	531.905	0.113	0.848	3	8	6	10	5.78	7.88	6.39	0.10	6.85

(continued.)

Table 2. Continued

Assessment Factor	Fuzzy Delphi expert questionnaire reliability preliminary test reliability statistics.				Fuzzy Delphi questionnaire for expert's reliability preliminary test overall statistics								
	Scale Mean After Removal of Items	Scale Scale Standard Deviation After Removal of Items	Modified Total Item Correlation	Cronbach's Alpha After Removal of Items	Most Conservative Cognition Value	Most Optimistic Cognition Value			Geometric Mean		Assessment Value		Expert Consensus Value
					Min (CLi)	Max (CUi)	Min (OLi)	Max (OUi)	CMi	OMi	Single value	MiZi	Gi
Power-saving monitoring management system 28	443.71	549.905	−0.254	0.851	4	8	6	10	5.49	8.17	6.34	0.68	6.93
Adequately light-weight building materials 29	443.00	507.667	0.522	0.840	4	8	6	10	5.49	8.23	6.74	0.74	6.94
Convenient design for building facility maintenance	444.14	529.143	0.154	0.847	3	8	6	10	5.81	8.68	7.33	0.87	7.10
Durable building design	445.00	517.333	0.404	0.842	3	8	6	10	5.16	8.46	6.68	1.30	6.93
Durable building design	443.86	525.810	0.239	0.845	2	8	6	10	4.75	7.67	6.20	0.92	6.68
Construction waste reduction	443.86	505.810	0.492	0.840	3	7	7	10	5.03	8.47	7.04	3.44	6.75
Reuse of construction materials	443.14	528.476	0.254	0.845	3	7	5	10	5.14	7.83	6.45	0.69	6.21
Construction pollution prevention	443.29	524.238	0.277	0.845	3	7	7	10	5.17	8.08	6.46	2.91	6.63
Classroom sound-proofing	442.86	504.476	0.709	0.838	4	8	6	10	5.60	8.38	7.00	0.78	7.00
School building lighting and illumination quality	444.43	521.286	0.236	0.846	4	8	8	10	6.17	9.02	7.54	2.85	7.60
Adequate ventilation	442.86	523.476	0.385	0.843	4	8	7	10	6.23	9.05	7.28	1.82	7.54

Prevention of vibrations and electromagnetic waves	442.86	528.476	0.348	0.844	3	8	5	10	4.78	7.30	6.06	−0.48	6.25
Nontoxic green construction materials	442.57	518.286	0.621	0.841	3	8	7	10	5.66	8.74	7.26	2.08	7.43
Establishment of water-saving facilities and water saving management	444.71	539.905	0.006	0.849	3	8	6	10	5.52	8.56	7.10	1.04	7.02
Establishment of water resource recycling and reuse facilities	444.57	527.619	0.280	0.845	3	8	7	10	5.52	8.51	7.19	1.99	7.38
Drinking water quality monitoring	443.86	538.476	0.060	0.848	5	8	7	10	6.03	8.69	7.48	1.66	7.46
Establishment of resources and waste classification and recycling management	442.57	521.619	0.526	0.842	3	8	6	10	5.12	8.39	6.53	1.27	6.91
Centralization of refuse disposal and beautification of refuse collection fields	442.86	509.143	0.538	0.840	3	8	5	10	4.68	7.14	5.72	−0.54	6.18
Establishment of sewage processing facilities	444.00	524.667	0.194	0.847	4	8	7	10	5.89	8.74	6.96	1.85	7.45
Campus air quality monitoring	444.43	540.286	−0.007	0.849	5	8	7	10	5.84	8.70	7.10	1.86	7.44
Laboratory safety and toxic substance control and management	444.14	530.810	0.102	0.849	3	7	7	10	5.28	8.21	6.77	2.93	6.75
Total number of factors No. of factors selected		56 36			Selection percentage		64.29%			Assessment value	6.84	Threshold value	6.91

established according to the research objectives of researchers. In this study, to achieve objectivity, the research of Ishikawa et al. (1993) was referenced, and based on their research suggestions, threshold values and suggestion values were used for screening assessment factors. The values of assessment factors must be greater than the threshold values and the suggested values. The threshold value is set according to Gi, which represents the importance of expert consensus. A high Gi value denotes a high level of expert consensus and high importance (Table 2). The threshold for this study was the arithmetic mean of Gi, namely the expert consensus value, 6.91. The assessment value was the arithmetic mean of the single-value geometric mean of each evaluation item, and it represented the subjective cognition data provided by experts. In this study, the assessment value was 6.84. Therefore, in this study, assessment factor indicators had to exceed the threshold value 6.91 and the assessment value 6.84 to be selected. The screening results are shown in Table 2. In total, 20 assessment factors were deleted, and 36 assessment factors were selected (64.29%). The symbol had deleted the following factors: school location selection; building economy of scale; green transport; life-cycle assessment of the buildings on campus; participation of teachers, students, and the public; land restoration; light hazards; bio-pond establishment; soil conservation; air conditioning energy-saving; green procurement; reuse of old campus buildings; construction waste reduction; resource reuse; construction pollution prevention; prevention of vibrations and electromagnetic waves; centralization of refuse disposal and beautification of refuse collection fields; laboratory safety and tox ic substance control and management; digital envi ronment and learning; and media on campus envi ronmental protection.

4 RESULTS AND CONCLUSIONS

4.1 Low-carbon campus policy and management and the importance of lifestyle and education

To achieve green, low-carbon, energy-saving benefits for green schools, compared with promoting routine energy-saving and energy management, the aspects of green, low-carbon policies and management and lifestyle education in schools are considered to be far more critical for the promotion of environmental awareness among teachers and students on campus. Because the third-layer assessment factor weights and importance indicators reflected all decision-making experts' study of Taiwan green campus assessment mechanism construction, during the planning stage, prior comprehensive overall demand planning and environmental assessment were em phasized.

4.2 Green overall construction design thinking in schools and communities

Overall, experts believe that campus environ ments have intangible effects on learners, who form views and opinions in relation to their sur roundings and are encouraged to address their eco logical concerns in other contexts. In schools, the teaching and administration are related to the space itself, driving the green school plan as a whole. Therefore, the assessment of campus construction and greening esthetics is a critical green school in dicator. In short, for the construction of green, low-carbon campuses for junior high schools and elementary schools, campus greening, favorable energy-saving facilities, and the planning of a diverse eco-campus and a healthy environment are indispensable.

REFERENCES

AASHE. Sustainable campus index. Available online: https://www.aashe.org/wp-content/uploads/2019/08/SCI-2019-Updated.pdf (accessed on May 22, 2019).

C. Ghent, A. Trauth-Nare, K., Dell, S. Haines, 2014. The influence of a statewide Green School Initiative on student achievement in K-12 classrooms. Appl. Environ. Educ. Commun. 13, 250–260.

A. Ishikawa, M. Amagass, T. Shiga, G. Tomizawa, R.Tatsuta, H. Mieno, 1993. The max-min Delphi method and fuzzy Delphi method via fuzzy integration. Fuzzy Sets Syst. 55, 241–253.

C.J. Kibert, 2012. Sustainable Construction: Green Building Design and Delivery; New York: John Wiley & Sons.

X. Luo, C. Ma, J. Ge, 2020. Evaluation model and strategy for selecting carbon reduction technology for campus buildings in primary and middle schools in the Yangtze River Delta Region, China. Sustainability. 12, 534.

H. Meiboudi, A. Lahijanian, S.M. Shobeiri, S.A. Jozi, R. Azizinezhad, 2018. Development of a new rating system for existing green schools in Iran. J. Clean. Prod. 188, 136–143.

N.H. Ramli, M.H. Masri, M. Zafrullah, H.M. Taib, N.A. Hamid, 2012. A comparative study of green school guidelines. Procedia Soc. Behav. Sci. 50, 462–471.

J.B.M.B. Sanfo, 2020. A three-level hierarchical linear model analysis of the effect of school principals' factors on primary school students' learning achievements in Burkina Faso. Int. J. Educ. Res. 2020, 100, 101531.

T. Savelyeva, J. Park, 2012. Complexity of campus sustainability discourse. In Sustainable Development at Universities: New Horizons; Filho, W.L., Ed.; Oxford, Peter Lang Scientific Publishers: Frankfurt, USA. 183–192.

Y. Shih, K. Liu, C. Chiang, N. Chen, 2014. Discussions on practice and effect of transforming idle spaces with energy-resource techniques in Taiwanese schools—Using elementary/secondary schools as an example. Energy Build. 68, 660–670.

C.M. Tang, 2014. New Perspective of Campus Planning; WuNan Publisher: Taipei, Taiwan.

J.C. Wang, 2016. A study on the energy performance of school buildings in Taiwan. Energy Build. 133, 810–822.

J.C. Wang, K.T. Huang, M.Y. Ko, 2019. Using the fuzzy Delphi method to study the construction needs of an elementary campus and achieve sustainability. Sustainability. 11, 6852.

S.C. Yeh, J.Y. Huang, H.C. Yu, 2017. Analysis of energy literacy and misconceptions of junior high students in Taiwan. Sustainability. 9, 423.

System Innovation in a Post-Pandemic World – Kin-Tak Lam et al. (Eds)
© 2022 Copyright the Author(s), ISBN: 978-1-032-24392-4

A narrative research on the career development history of an artist

Artde Donald Kin-Tak Lam*
School of Design, Fujian University of Technology, Fuzhou, Fujian Province, P.R. China

Vivien Yi-Chun Chen
Fujian University of Technology, Fuzhou, Fujian Province, P.R. China

Con-Rong Wang
JiMei University, Xiamen, Fujian Province, P.R. China

ABSTRACT: This study mainly proposes a method to explore the career development history of an artist by using the arched model theory. First of all, the narrative research method is used to collect, analyze, code, and summarize the data, so as to deepen the psychological world of the research object and achieve the best understanding and description. Then, based on the theory of "Arch Model" of career development, this paper demonstrates how it affects their personal career development process and career development tasks. Through the analysis and results of the above career development, we can understand the elements of an artist's success. The results of this study provide a theoretical reference for people to achieve "career maturity" and "self realization" in career development theory, and provide suggestions for students with uncertain career choices or people in employment, so as to understand their career tasks.

1 INTRODUCTION

Everyone has his own deep-seated thoughts or unrealized regrets. In the process of their creation, artists often invest in their life experience, so that their works also reflect their inner thoughts and life experience, especially the life connotation reflected in their works at different stages. Therefore, studying the past life experience of artists through the works they created in different stages or periods is an important basis for discussing their career development and understanding the creative process.

Art has no correct way of expression and no so-called fixed way of expression. As an art reflecting the image of life, it should be extremely rich and complex, and artistic creation has not been brought into full play. Finally, it cannot advance. Therefore, artists should pursue perfection and never stop the performance of their artistic life. If an artist lacks enthusiasm and love for art, he will not bother to think and explore.

A ceramic artist adheres to the creation of art, not the mainstream of market orientation. His sincere and simple personal characteristics can feel that artistic creation has a special significance and even a mission for him. The ceramic artist's art growth process and art teaching are a very important part of his personal career development. The artistic works and creative style of the research object not only show the deep

thinking of the research object, but also show the research object's views and attitudes toward social culture.

Art works are the result of the combination of the artist's internal and external presentation. This study constructs and analyzes the artist's career process through an artist's story of his life experience and his understanding of each artistic creation idea, so as to understand his process of life practice and how to create self-identity career development and artistic achievements. Therefore, based on the narration of the life experience of this artist, this paper studies the career development stage of a craftsman, which is an important basis for understanding the connotation of the career development history of an artist and affecting the presentation level of artistic creation.

2 LITERATURE

The artist's creation is through the cultivation of the soul of the author and the nurturing of feelings. If the deliberate artist wants to run the business, he will create no artistic works with vitality. The artist's creative process is an accumulation of natural evolution and life. It is an indescribable and continuous expansion state. The artist's career development history can bring up his soul, activate his thinking, and form his unique artistic style through interaction and self-dialogue with his surroundings. Therefore, by exploring the artist's

*Corresponding Author

DOI 10.1201/9781003278474-19

career development history, we can find his insistence on creation and amazing moving.

Before the 1950s, the predecessor of career development was "vocational development," which emphasized the importance of the combination of career and itself. The study of career development theory mainly rose in the 1950s, based on individualism and developmental psychology, emphasizing the process of life. The research on career development began from the research on career development and career psychology engaged by Super (1957a, 1975b). Super (1957a, 1975b) propose 12 propositions of career and career development, which is a lot of research arguments formed by integrating the views of many scholars and the basis of his long-term research and injecting the traditional career into new ideas.

In the study of professional behavior in Super et al. (1990), "self-concept" is a very important part. It refers to that people consider their own characteristics and the characteristics required by their career at the same time. The "professional self-concept" is a part of the overall self-concept. It is developed through life, psychological growth, observation of work, recognition of others in work, environment, and experience. The purpose is to assimilate the differences and similarities between themselves and others. The deepening and broadening of cognition and experience of the working world and the formation of more complex career concepts are the driving force for individuals to follow the career form in their life. Self-concept is the core of Super's career development theory. Super et al. (1990) derived the "Arch Model theory from the self-concept theory, in which the upper part of the arch represents the individual self, and the left column and the right column are the physiological cornerstone and the geographical cornerstone, respectively. Figure 1 shows the Super's "Arch Model," which is explained as follows:

(1) The physiological cornerstone of the arch supports psychological traits. There are factors such as personal needs, wisdom, price and quality, sexual orientation, and interest, which constitute personality variables and affect personal achievement performance.
(2) The geographical cornerstone of the arch supports factors such as community, economy, society, family, school, peer group, and labor market, and affects social policies and employment standards.
(3) The arched synthesis of the two cornerstones, describes the variability of different roles experienced in one's life, which is influenced by the stage of career development and self-concept, and is related to one's career choice and development.

In 1990, Super revised the developmental task and indicated that age and career transition are very flexible and do not necessarily occur according to the career development discipline. He thinks that the completion of tasks at all stages is "Vocational Maturity." The characteristics of vocational maturity include planning for the future, taking responsibility, awareness of career to varying degrees, etc. In other words,

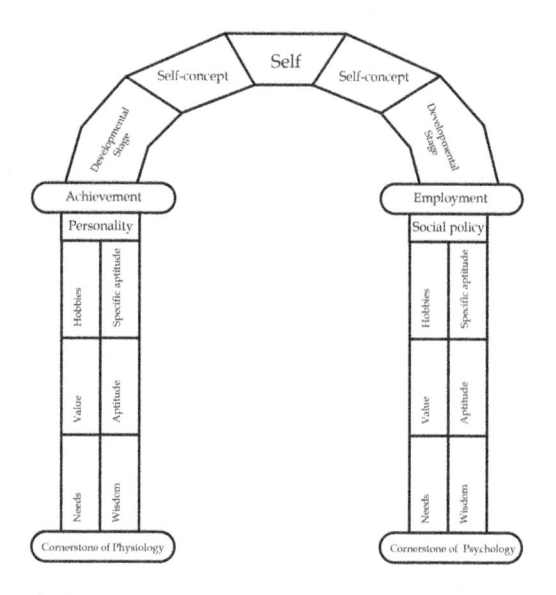

Figure 1. Arch model of Super's career development.

"Vocational Maturity" can also be said to be the process of continuous realization of self-concept. This argument can be used as a reference for vocational training and career counseling, and is also quite helpful to the development of career counseling and vocational education programs. McMahon (2005) discussed the practical application of the Systems Theory Framework of career development as a guide to career counseling. Giannantonio and Hurley-Hanson (2006) introduced the influence of image norms in various stages of Super's career, which is of great significance to individuals, organizations, and career consulting professionals. Tejedor (2007) developed a program of vocational counseling that allows the participants, students of Secondary Education, to improve their vocational maturity taking into account their level of vocational self-efficacy. Hoekstra (2011) indicated that combining different career roles predicts success as well as satisfaction. Liu et al. (2017) explored the relationships between vocational interests and career maturity in the context of China. Kantamneni et al. (2017) provided psychologists and counselors with a more nuanced understanding of how career decisions are made for Asian American college students. Chen (2020) purposed how social support, career beliefs, and career self-efficacy contribute to college athletes' career development. The results of this study indicated that career beliefs positively contributed to the prediction of career self-efficacy and career development. Hsu et al. (2021) examined how young people build their employability through self-regulated learning processes and whether career maturity might be a critical antecedent of self-regulated learning processes. The results of this study from 257 undergraduates in Taiwan indicated that both metacognitive and motivational self-regulated learning contribute to young people's perceived employability.

3 RESEARCH PROCESS AND ANALYSIS

In this study, qualitative research is closely related to certain research methods and processes. In order to achieve the best understanding and description, we need to adopt a historical perspective. Semi-structured interviews were conducted to allow researchers to walk into the world of individual existence by giving respondents the openness to ask questions. Let the interviewees tell their personal stories, the speaker leads the listeners into their world, and the interviewees narrate their specific experiences in life, so as to construct an integrated self again. Thus, this study invited interviewees to describe the growth process and creative process, and guided the researchers to organize the research results into a text presentation.

Based on the respect and protection of the rights and privacy of the respondents participating in this study, the researcher will prepare the interview consent form in advance before the interview. The respondents are asked to read the content of the interview consent form in detail, understand the research purpose and process, abide by the research ethics, and ask the respondents to sign their consent. The interview can be conducted only after the researchers and the respondents reach a consensus.

The interview outline of this study is a semi-structured interview outline based on Super (1957a, 1975b) career development theory, combined with the research objectives and problems. The main purpose of the interview outline is to guide the direction of the interview. The design content of the interview outline is particularly important to make the interviewees speak more richly and fluently. The research process is as follows:

(1) Preliminary operation: (a) determine research objectives and (b) enter the research field and participate in activities.
(2) Leading research: (a) draw up an interview outline; (b) leading conversation; (c) analysis of leading conversation; and (d) revised interview outline.
(3) Formal interview: (a) obtain mutual consensus between researchers and respondents to determine the conduct of formal research; (b) understand the personal artistic creation of the interviewees and start to enter the formal interview; and (c) record data, edit data (transcription), and reconstruct reality.
(4) Collation of research results: (a) inspection and verification of research data; and (b) writing research report.

According to the above process, the analysis and collation of data are described as follows:

(1) The interview content is transferred verbatim.
(2) The interview content is encoded verbatim.
(3) Coding the interview content according to Super's "Arch Model."
(4) Write the interviewee's career story and artistic creation story.

4 RESULTS AND DISCUSSION

This study takes an artist as the research object (interviewee) and through the above research process, leads the researcher into the interviewee's world, deeply understands his growth process, and witnesses that he has created a satisfactory career with his own willpower and persistence. In his whole career, he had a smooth career course, with the help of friends, and of course, he also encountered difficulties and crises. Therefore, based on the interview results of the respondents, sort out their career stories, understand how the respondents are prepared, and how to break through the difficulties in the face of self-development, so as to strive for better career development:

(1) Self-processing and affirmation of interviewee: (a) when the interviewee realized that the study environment was contrary to his expected career direction in the future, he stopped learning in school, settled his mood, and thought about his future career; and (b) the interviewees returned to school to continue their study after a period of time in society, but still determined that this study was not suitable for their own career development, and gave up further study again to become the spiritual pillar of the interviewees' constant interest (writing and painting creation).
(2) Rally forces again: (a) the interviewees have successively employed in different places, learned skills in their spare time, wrote many articles, and created many paintings; (b) the interviewee returned to his hometown to settle down after getting married and having children, and once opened a mounting shop, but closed down because of poor business; (c) the interviewee sorted out the daytime painting works over the years and resolutely launched the first daytime painting exhibition in their life with the help of friends.
(3) Clarify the solvable and unsolvable problems of life: the interviewee will be confident that he can do his creation well and let others complete the other parts.
(4) Self-awareness: the interviewees transformed their mind and emotions into artistic creation through creation, found the explosive power to achieve themselves in the creative process, and the creative inspiration was endless.
(5) Everything is done without any dazzling achievements: the interviewee insisted on his unique creative style, and his works did not flow into form and appeared highly original, because the interviewee only wanted to express his ideas, which created the interviewee's continuous artistic career.

5 CONCLUSION

In this paper, we study the career development of artists, and discuss their personal career process and career development achievements according to Super's career development theory and arch model. The research results understand the growth process of artists and witness that artists create a happier career by relying on their own willpower and persistence in interest.

(1) Self-processing: affirm yourself and be brave to try many times.
(2) Rally forces again: be clear about your goals and understand that optimism will save you from danger.
(3) Clarify the solvable and unsolvable problems of life: regard life as a unit and clarify what you are good at and what you are not good at.
(4) Self-awareness: the ability to make good use of adversity pressure can transform itself into self-help.
(5) Everything is done without any dazzling achievements: the respondents did not create applause for the public, only to make their lives bright.

ACKNOWLEDGMENT

This research is funded by Fujian Province Social Science Project No. GY-S19045 & GY-S20050, and Fujian University of Technology Project No. GY-S18014.

REFERENCES

C.C. Chen, 2020. *J. Hosp. Leis. Sport To.* 26 100232.

C.M. Giannantonio, A.E. Hurley-Hanson, 2006. *Career Dev. Q.* 54(4) 318–330.

H.A. Hoekstra, 2011. *J. Vocat. Behav.* 78(2) 159–173.

A.J.C. Hsu, M.Y.C. Chen, N.F. Shin, 2021. *Int. J. Educ. Vocat. Gui.* DOI: 10.1007/s10775-021-09486-z.

N. Kantamneni, K. Dharmalingam, G. Orley, S.K. Kanagaingam, 2018. *J. Career Assessment.* 26(4) 649–665.

Y. Liu, K.Z. Peng, Y. Mao, C.S. Wong, 2017. *J. Career Dev.* 44(5) 425–439.

M. McMahon, 2005. *J. Employment Couns.* 42(1) 29–38.

D. Super, (1957a). *The Psychology of Careers: an Introduction to Vocational Development.* New York: Harper.

D. Super, (1957b). *Vocational Development: A Framework for Research.* New York: Teachers College, Columbia University.

D. Super, L. Mark, M. Charles, 1990, in *Career Choice and Development*, eds D. Brown & L. Brooks, 167–261, San Francisco, CA: Jossey-Bass.

E.M. Tejedor, 2007. *Rev. Psicodidact.* 12(1) 121–129.

System Innovation in a Post-Pandemic World – Kin-Tak Lam et al. (Eds)
© 2022 Copyright the Author(s), ISBN: 978-1-032-24392-4

Study on the combination of augmented reality technology and children's folk game—an example of the world toy exhibition in Yunlin Re-toy House

Chia-fang, Hsu*
Department of Early Childhood Educare, Transworld University of Technology, Douliu, Yunlin, Taiwan

Liang-Yin, Kuo
Smart Machinery and Intelligent Manufacturing Research Center, National Formosa University, Hu-Wei, Yunlin, Taiwan

Chien-lung, Chiu
Department of Visual Communication Design, Transworld University of Technology, Douliu, Yunlin, Taiwan

ABSTRACT: Children's folk game reveals the transformation of lifestyles and progression of culture of people through time. It is an output of the wisdom of time, and also an important treasure needed to be inherited, collected, and introduced to younger generations. At the heart of the collection and introduction of children's folk games lies not only the dimension of "play" but also helping players to look into the depth of the cultural meaning of the game. In this study, more than 40 children's folk games played worldwide are collected for a world toy exhibition, integrated with the application of augmented reality (AR) technology. The purpose of using AR is to digitize the full picture of the games for long-time preservation as well as giving new looks to old games. This study builds up the AR materials for each game consisting of videos and films, which are easy for children to operate; moreover, the integration of AR with modern interactive technology benefits the promotion of the exhibition as it attracts visitors' interests and draws children's attention to exploring the games.

Keywords: Children's folk game Augmented Reality Technology

1   INTRODUCTION

1.1   *Folk game*

Playing is an integral part of human growth, as children learn basic life skills through playing. Children's folk games are not just games to be played, but they are also part of the history, which relates to age-old social values, beliefs, religions, and customs, and thus represent an output of the wisdom of time. Samovar, Porter and McDaniel (2012) defined children's folk games as cultural symbols that can enable a culture to preserve what is important and worthy of transmission. Children's folk games are valuable assets, and, thus, it is important for us to learn, share, preserve, and transfer the games to younger generations. For the reason stated above, a World Toy exhibition was curated at Yunlin Re-toys House with the application of the augmented reality (AR) technology, aiming at digitizing the full picture of the games for long-time preservation as well as giving new looks to old games.

1.2   *Augmented reality*

AR is an enhanced digital version of the real world, which is achieved through the application of digital visual elements, sound, or other sensory stimuli delivered via technology. In recent years, the introduction of AR skills into exhibitions or the game industry has been on the rise. For example, Pokémon Go, the game where users need to "catch" Pokémon hiding in the world around them. Animated creatures are superimposed onto what players can see through their device's camera. In Taiwan, Yi-lan International Children's Folklore & Folkgame Festival, one of the most famous thematic festivals, has integrated AR technology with the toys and exhibitions to a great extent since long. In the context of a museum, the goal of exhibition is to entertain as well as educate the audience, and the integration of AR skills and related interactive technology can help enhance the visitor's knowledge, make learning more fun, and encourage more involvement by bringing forward information, liveliness, and dynamism to the exhibited objects. There are many possibilities with the use of AR in exhibitions. The most straightforward way is to use it to add explanations of items in exhibition. As a result, in this study, taking the example of the World Toy exhibition, the researcher utilized AR skills to build an app consisting of 36 classical children's folk games collected worldwide for enhanced experiences, education, and preservation.

*Corresponding Author

DOI 10.1201/9781003278474-20

2 EXPERIMENTAL SECTION

The main purpose of the AR app designed in this study is to develop a knowledge-based and preservable touring guide as well as to explain the exhibited items to visitors. It is expected that the AR app could be integrated with films, texts, and interactive elements such as animated images, which could make learning more fun. To develop the app, first, the content of the AR app was designed, the process of AR production was analyzed, and the Unity version 2021 working system and Vuforia framework were employed.

Then, the films about the historical background and introduction of each folk game were video-shoot and then produced. Following that, the 2D materials, including the cards and animated images to be used in the app, were created. After developing the 2D materials, the image recognition and tracking were run in Vuforia, and after integrating the app with all the films and materials, it was edited and built through the Unity working system. Finally, the AR app was subjected to tests via smartphone or electronic devices followed by implementing further modifications, if required, before it was launched.

3 RESULTS AND DISCUSSION

This section consists of three parts: the introduction to the AR app production process, the AR images, and the AR app implemented in the exhibition.

3.1 *The AR script process*

In this study, 36 folk games were collected and digitized using the AR app. Figure 1 shows the flowchart representing the process of AR app implementation:

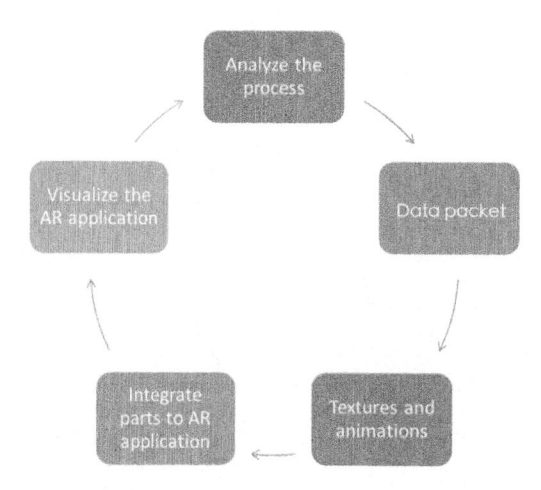

Figure 1. The process of AR app implementation.

3.2 *The AR app*

Based on the above-mentioned process, the films, depicting the origin and method of playing every folk game, were shoot as the key content of the AR app, as shown in Figures 2 and 3.

Figure 2. Film shooting of folk games.

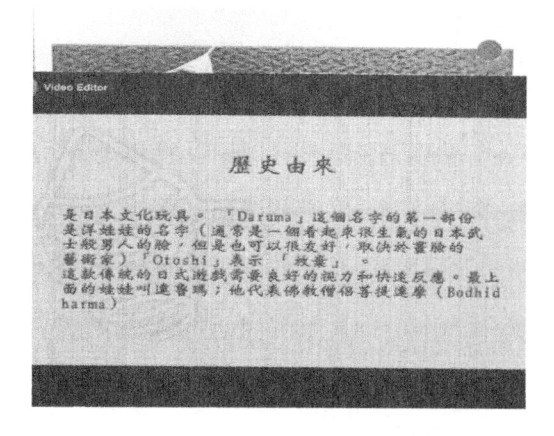

Figure 3. Film shooting of the origin of folk games.

After completing the film shooting, the 2D image production, including the image cards of folk games as well as the animation of images to be displayed in the apps, was accomplished, as shown in Figure 4.

The final stage was to integrate all the materials with the AR app and then visualize the performance of the AR app.

The app can be installed in the smartphone or tablet, and users can view the content by scanning the image card of each game. On opening the app, an introduction about the origin and historical background of the folk games followed by films demonstrating the method of playing the game will show up in sequence. Such an arrangement would help visitors easily visualize the game pieces and understand them. An animated image of the game piece would pop up on the top of the screen with movements in different dimensions to enhance the experience. Figures 5–7 demonstrates a

Japanese folk game "Daruma" as an example to show how the AR app will work.

Figure 4.   Image card of folk games.

Figure 5.   Animated image popped up on the top of the screen.

Figure 6.   Introduction of the game.

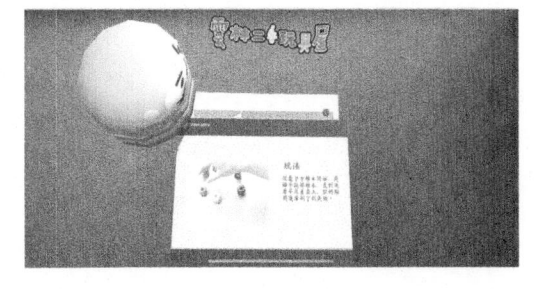

Figure 7.   Film demonstrating the game.

Figure 8.   AR zone.

Figure 9.   Visitors exploring the AR app.

3.3   *AR zone in the exhibition*

During the World Toy exhibition, zone 2 was designated for the display of the AR app followed by zone 3, where the visitors actually see and try the folk game pieces. A total of three tablets with the AR app installed and three sets of folk game image cards were provided in zone 2. Visitors were free to take any image card they were interested in and scan it for an introduction of the game. Figures 8 and 9 show how visitors self-learned using the app in the AR zone.

4   CONCLUSION

Over the past few years, AR has gained immense popularity and has been developed for use in museums. In this paper, a case of how AR skills can be integrated with exhibitions is presented. In the World Toy exhibition, many advantages of using the AR app have been demonstrated, including providing easy access to knowledge, reducing working staff, attracting the visitor's interests to the exhibition, and motivating children and old people to explore the app. The visitors' perceptions and satisfaction toward utilizing AR, and the advantages of integrating AR technology with exhibitions need to be further investigated. For the future study, it is suggested that more varieties of AR or interactive technologies can be developed and incorporated in museum exhibitions.

REFERENCES

Gillam, Scott. "Spotlight VR/AR: Innovation in Transformative Storytelling – MW17: Museums and the Web 2017," February 28, 2017. https://mw17.mwconf.org/paper/spotlight-vrar-innovation-in-transformative-storytelling/

Haugstvedt, A. C., and J. Krogstie. "Mobile Augmented Reality for Cultural Heritage: A Technology Acceptance Study." In 2012 IEEE International Symposium on Mixed and Augmented Reality (ISMAR), 247–55, 2012. https://doi.org/10.1109/ISMAR.2012.6402563.

Prabhu, Sanket. "Types of Augmented Reality (for Me and My Business) – ARreverie Technology," December 18, 2017. http://www.arreverie.com/blogs/types-of-augmented.reality/

SAMOVAR, L. A., PORTER, R. E., MCDANIEL, E. R.: Communication Between Cultures (8th edition). Belmont, CA : Wadsworth, Cengage Learning, 2012.

Yoon, Susan A., Karen Elinich, Joyce Wang, Christopher Steinmeier, and Sean Tucker. "Using Augmented Reality and Knowledge-Building Scaffolds to Improve Learning in a Science Museum." International Journal of Computer-Supported Collaborative Learning 7, no. 4 (December 2012): 519–41. https://doi.org/10.1007/s11412-012-9156-x

Yoon, Susan A., and Joyce Wang. "Making the Invisible Visible in Science Museums Through Augmented Reality Devices." TechTrends 58, no. 1 (January 1, 2014): 49–55. https://doi.org/10.1007/s11528-013-0720-7

System Innovation in a Post-Pandemic World – Kin-Tak Lam et al. (Eds)
© 2022 Copyright the Author(s), ISBN: 978-1-032-24392-4

Tacit knowledge in digital design: From master to student

Jia-Xiang Chai*

Guangdong Literature & Art Vocational College, Guangzhou, Guangdong, China

ABSTRACT: This paper discusses research that employed practice-led and action research methods to study the tacit knowledge of digital design practice and its application to teaching. Polanyi's theory of tacit knowledge is used to analyze the nonverbal, experience-based knowledge of masters and students to construct a discursive relationship between the practices of design and teaching. This study implemented an 8-week interface design program. Within the setting of the design studio, a series of the student's design processes/actions were created and documented. By analyzing the crucial and subsidiary knowledge of the design processes, several distinct patterns of action and thinking emerged. These patterns were synthesized into four modes of thinking, which integrate the mind, programs, skills, and materials. The outcome of the study is a preliminary model that describes digital design as a dynamic multimodal thinking process, integrating visual perception, digital technique, material actions, and expressive ways of thinking. The discussion includes a detailed description of the research methods, the data analysis, and the embodied nature of cognition in the digital design process.

Keywords: Tacit knowledge, Digital design process, Design thinking, Visual thinking, Computational thinking, Material thinking

1   INTRODUCTION

Digital design is a profession that integrates art, design, and computer science, in close relation with illustration, typography, animation, and music, among others. At the same time, it is a unique, digital sector (Özcan and Akarun, 2002a). However, when doing the graduation design project, many digital media design majors from colleges and universities in China would copy some classic demonstrations rather than work through the problems themselves. They have not transferred their learning into works. Based on Polanyi's (1966) theory of tacit knowledge, through a practical course in an interactive design studio, the author analyzes the actions and thinking of teachers and students that could affect the problem-solving skills of students.

2   LITERATURE REVIEW

2.1   *Theory framework: tacit knowledge*

Polanyi (1958) mentioned that, "There are two kinds of human knowledge, explicit knowledge (i.e. to 'know that' something is the case) and tacit knowledge (i.e. "know how" to doing things)." In *The Tacit Dimension*, he emphasized that tacit knowledge could be further divided into two categories: one is skill, technique, and art (i.e. weak tacit knowledge) and the

other is judgment, connoisseurship, interest, and creativity (i.e. strong tacit knowledge, or tacit ability) (Adloff et al., 2015; Polanyi, 1983). Tacit ability depends on talent and practice. For beginners, the ability can be developed via rules and examples, while for advanced learners, demonstrations are more productive than precepts when they are trained on skills that can hardly be explained with words (Polanyi, 1958;; Schindler, 2015; Yu, 2002). According to the tacit epistemology, "learning by example" helps enhance weak tacit knowledge, and analogical reasoning elevates strong tacit knowledge. The transfer of tacit knowledge requires group practice, demonstration, or exchanges in a common space.

2.2   *Design education and tacit knowledge*

As with all design majors, "learning by doing" (Dorst & Reymen, 2004) is the most common approach in digital design teaching (Smith, 2015). In a design studio, teachers very often use design programs to carry out teaching activities (Lawson & Dorst, 2009), and encourage the strategic use of design methodologies, such as a situational story and role playing (Curry, 2014)

In the research on the design process and tacit knowledge, Lin (2002) developed a framework of the knowledge transfer process within design teams based on the SECI model (Nonaka & Takeuhi 1995). Wong

---

*Corresponding Author

DOI 10.1201/9781003278474-21

et al. (2016) referred to Lin's findings when exploring the growth path of designers and the knowledge transfer among them in the project.

3 RESEARCH METHOD

Qualitative methods of practice-led digital media design research and practical action research were employed. The teaching practice was designed as part of an 8-week interactive media design course, with the subjects being 24 junior students who did internships at the design studio. The students have a professional background in digital media design and a related project experience. The theme of the course was "'Digital Expression of Silk Figured Lions in Foshan." Students worked in groups to design mobile-based interactive works, which would help promote the manufacturing process of the intangible cultural heritage of silk figured lions of Foshan.

The course was jointly undertaken by a design teacher with 8-years of teaching experience and a inheritor of the intangible cultural heritage. To figure out how students gain tacit knowledge when working in the studio, the author did both action research and de- sign research based on the project cycle of planning, implementation, and reflection (Lawson & Dorst, 2009). With the consent of the research subjects, pho- tos were taken to record some of their special words and actions. After 8 weeks, our observers (two researchers trained in cultural studies and ethnographic research) conducted participatory observation and recording.

In the early stages of the course, the inheritor organized workshops, demonstrating the four steps of silk figured lion making to students, passing their ideas on the color silk techniques and the lion dancing culture down the line, and introducing their representative works. In addition, they released design themes and requirements of the client. At the end of the course, the inheritor graded students' works based on design goals and innovativeness. The design teacher from the studio guided students in doing practice-oriented project research, implementation, and testing. Gaining complementary skills, the students worked in a group of six, discussing with the teachers on the design concept and the software proficiency, painting skills, and de- sign basics required. Then, they conducted some research on prototyping, low-fidelity testing, and high-fidelity testing. At the end of the course, the works were reviewed by professional designers, inheritors, and end-users in terms of product usability, ease of use, accessibility, and user satisfaction (Carson, Peterson, & Higgins£¬2005).

4 DATA ANALYSIS AND DISCUSSION

The aim of this article is to find out how the tacit knowledge on digital design is transferred between teachers and students, and to figure out their actions and thinking, the author asked the students and teachers to explain (a) "what they are doing" and (b) "why they are doing it," followed by describing, reasoning, and analyzing all their actions and ideas (Schindler, 2015), which showed that students were able to clearly and explicitly convey knowledge (Smith, 2001). This helped uncover all their actions, ideas, reasoning and analytical skills (Schindler, 2015), and knowledge communication abilities (Smith, 2001). By doing so, the author figured out the crucial actions and thinking involved in the process. In addition, the tacit knowledge transfer among all participants was recorded based on Nonaka's (1991) and Lin's (2002) frameworks, as shown in Figure 1.

Figure 1 also shows the flow of actions, explicit presentations, and tacit knowledge transfer between teachers and students. In the "students' actions" column, the actions are coded and summarized based on tacit epistemology, SECI, and related research results (Nonaka & Takeuhi 1995; Lin, 2002; Marinkovic, 2021). When learning about the silk figured lion making techniques, the students follow the inheritor to clean and swab the bamboo, and for learning computer basics, the students regulate parameters under the instruction of teachers. These are all labeled "imitation." In the "teachers' actions" column, the actions (demonstrations, rule briefing, and analogical reasoning) are coded and summarized based on tacit epistemology, SECI, and related research findings (Yu, 2008, 2012). Moreover, teachers' actions are further classified and described based on what they do in the real world and in the digital world. For example, the inheritor taught bamboo peeling by showing the process. When she shared the digital projects she had done, she used a 3D printed model. Therefore, all the inheritor's actions take place in the real world. The actions are classified in this way because it helps the students to distinguish the digital/virtual objects from the real ones (Smith, 2015).

The "themes of knowing" column covers four areas, namely project background (manufacturing techniques of silk figured lions), computer technology, visual animation, and user-centered design knowledge (such as the design process and digital design specifications). These are the areas commonly covered by the interactive design courses proposed by Özcan and Akarun (2002). During the teaching process, the teachers transfer students both the nonverbal or material-based knowledge, and the verbal and/or conceptual knowledge. Based on these themes, a total of four modes of thinking are proposed, namely material thinking (MT), computational thinking (CT), visual thinking (VT), and design thinking (DT), as shown in the first column of Figure 1.

The arrows in Figure 1 show the direction of knowledge flow. The explicit and tacit knowledge un- der the four cognitive themes flows from the teachers to the students. As the students work in groups at the studio, the knowledge flows upward in spiral among students, as described in the SECI model. It is an upward spiral. When interviewed, the students mentioned that they

the modes of thinking	themes of knowing	teacher's action		students' actions
		in real world	**in digital world**	
Material thinking	The knowledge of project background about silk figured lions of Foshan	Demonstration (example reasoning) • Bamboo cleaning • Bamboo grilling • Paper wrapping • Swabbing; •		Observation / Experience / Imitation
		Rule Brief • Material selection; • Classification of lion images; • Requirements of decoration; •		
		Analogical Reasoning • Lion dancing culture inheritance • Presentation of digital works;		Combination
computational thinking	computational technology	Rule Brief • Development process brief	• Development process demonstration • Programming language • Work code adjustment • Parameter adjustment • Work debugging	Recording / Analogical reasoning / Transfer
		Analogical Reasoning	• Revision to some case templates • modification open source	Computing / Imitation / Combination / Innovation
Visual thinking	Arts/Animation	Demonstration (example reasoning) • Composition adjustment • Animated simulation	• Abstraction of images and patterns • CG demonstration	Sorting / Experience
		Analogical Reasoning • Visual style similarity brief		Imitation & practice / Transfer / Innovation — Draft / Discussion
		Rule Instruction & Demonstration • Object movement rule	• PS, AI, AE and Spine introduction;	Metaphor / Recording
Design thinking	User-Centered Design	Analogical Reasoning • Presentation of digital works	• Design process adjustment • Interactive prototype making demonstration	Observation / Investigation / Imitation / Sorting — Analogical reasoning / Similarity identification
		Demonstration (example reasoning) • Demonstration of investigations • Demonstration of using scenario acquisition		Summarization / Combination / Transfer — Metaphor / Draft / Discussion / systematists

Inheritor & design teacher on digital design from studio

students Group

Figure 1. Date analysis chart mapping tacit knowledge functional in a design studio.

GroupA First A+	GroupB Second A	GroupC Third B+	GroupD Fourth C

Figure 2. The works of students.

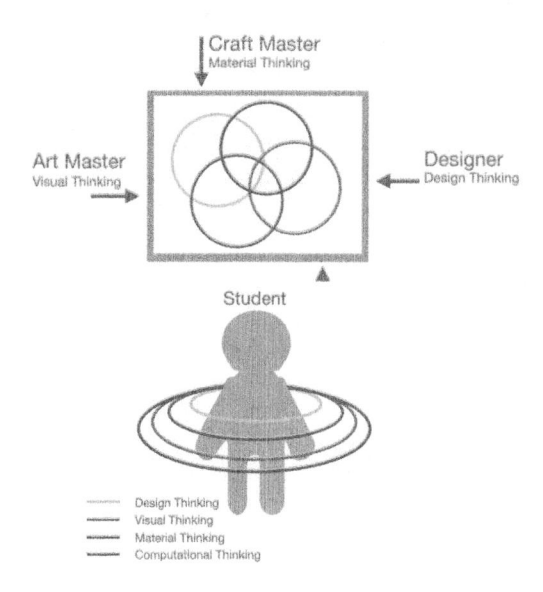

Figure 3. Four modes of thinking entangled in the digital designer thinking model.

learnt from each other in the group and via the progress report.

The relationship between thinking and action is the one between nonverbal behavior and explicit con cepts, with the bridge between them being the final works of the students. After being reviewed by the in- heritor, design teacher, and target users of the works, Group A got the highest score. The judges agreed that the work was designed with the aim of solving actual problems, although it imitated the visual design and interactions of other works. Groups B, C, and D occupied the second, third, and fourth places, respectively. According to the field notes and coded data, students in Group A were outstanding in terms of DT, especially in similarity identification, metaphor using, and systematisms. During the interview, Group A students admitted that they spent much time imitating the design teacher to "capture the usage scenarios of the work" via role playing see Figure 2).

Students in Group B performed well overall in terms of the cognitive themes and the thinking modes. This was reflected in program stability, visual effects, and interactive performance of the work. Compared with those from Groups C and D, these students did a much better job in thinking and actions. But, when compared with students of Group A, they were weaker in terms of DT. Students in Group B reported that they had spent much time on visual animation and mastering computer skills, and that they benefited much from the on-site demonstrations of the inheritor.

The work of Group C did solve actual problems, but was similar to the demonstration case in terms of visual effects and interactions. The field notes showed that students in Group C were much weaker than those of Groups A and B in innovation, analogic reasoning, and knowledge transfer. Four students in Group C reported

that their thinking and creativity were restricted by the demonstration case.

Five of the six students in Group D did not show up at the inheritor's demonstration. They lagged far behind in terms of CT, VT, and DT. The students reported that they lacked imagination about concrete objects due to their absence from the demonstration session. The students who took part in the technique demonstration session could not express ideas clearly to other members. All these led to an unsatisfactory work. The field notes and coded materials show little flow of knowledge under the cognitive themes and thinking modes.

The thinking and actions of students and teachers are presented in Figure 1. Actions that took place when the students received the nonverbal actions and explicit concepts from the teachers were analyzed based on the students' works (Figure 2) The four modes shown in Figure 3 originated based on the thinking mode of the students majoring in digital media design and by combining the discussions of this pa per with existing findings on thinking mode.

5 CONCLUSIONS: DIGITAL MEDIA DESIGNER THINKING MODEL

The digital design process is a dynamic multi modal thinking process where the mind, body, and materials are seamlessly woven together (Marinkovic, 2021). Figure 3 shows the digital media designer's or students' thinking model. Among them, DT is the most crucial mode of thinking, followed by MT, CT, and VT. These modes of thinking are activated alternately throughout the process. When one mode of thinking becomes foreground awareness, the others would recede. The thinking mode and activities determine the quality of the final work of the digital media designers. As shown in Figure 3, the thinking modes are represented with four rings, functioning alter nately. The size of the rings depends on the changes in cognition as the students continue to gain more knowledge. No thinking mode shall be neglected; otherwise, the creative design process could be dis torted.

Based on the observation and study of this interactive design course in a design studio, it is found that one action could take place under all thinking modes. One thinking mode in the students is activated based on the cognitive theme. As DT is the first and most important mode, actions under this mode have a direct impact on whether a work solves actual problems. As there is causality involved between the designers and the materials (Bolt, 2004), the interaction between the designers and the objects leads to insight, and MT is the empirical knowledge gained through practice, experiment, and trial and error. This thinking is an important and active mediator in the human's cognitive development (Smith, 2014; Marinkovic, 2021; Kampen, 2019). As reported by students in Group D, their lack of multidimensional MT affected their work design. In demystifying CT, Shute et al. (2017) noted that CT covers six aspects, namely decomposition, abstraction, algorithm design, debugging, iteration, and generalization. It is a computer engineering thinking mode, and an essential digital competence in the 21st century, particularly for a major of digital media design (Wing, 2006, Shute et al., 2017). For digital media designers, VT in the painting process is analytical, and functions as part of the concrete action–perception cycle. The clever use of VT is com plementary to the programming language under CT, i.e. false digital design (Arnheim, 1969. & Varela et al. 2016). The cross-application of CT and VT makes up for the deficiencies of DT as well (Bailey & Borwein, 2011). As shown in Figure 3, to reform the digital media design teaching in studios, the teachers need to be equipped with professional thinking and cognition, which is critical to fostering the thinking modes and actions of the students. As the explicit and tacit knowledge alternation increases, the students need to socialize, externalize, combine, and internalize the knowledge in studios, experiencing a spiral-shaped growth path.

This research was conducted by using qualitative methods, with the findings concluded in the form of the practice and field notes. More research can be conducted to quantify the actions of teachers and students, and on work evaluation, thereby providing more reliable support for the reform of digital design education and teaching.

REFERENCES

Adloff, F., Gerund, K., & Kaldewey, D. (2015). Locations, translations, and presentifications of tacit knowledge: An introduction. In F. Adloff, K. Gerund, & D. Kaldewey (Eds.), Revealing tacit knowledge: Embodiment and explication (pp. 7–17). Bielefeld, Germany: Transcript.

Bolt, B. (2004) Art Beyond Representation: The Performative Power of the Image. New York: I.B. Tauris.

Carson, S., Peterson, J. B., & Higgins, D. M. (2005). Reliability, validity, and factor structure of the creative achievement questionnaire. Creativity Research Journal, 17(1), 37e50.

Curry, T. (2014). A theoretical basis for recommending the use of design methodologies as teaching strategies in the design studio. Design Studies, 35(6), 632–646. doi:10.1016/j.destud.2014.04.003

Dorst, K., & Reymen, I. (2004). Levels of expertise in design education. In International Engineering and Product Design Education Conference, September 2–3, 2004, Delft, The Netherlands.

Kampen, S. (2019). An investigation into uncovering and understanding tacit knowledge in a First-Year design studio environment. The International Journal of Art & Design Education, 38(1), 34–46. https://doi.org/10.1111/jade.12171

Lawson, B., & Dorst, K. (2009). Design expertise. Oxford: Elsevier.

Lin, L. S. (2002) A Study of Performance Evaluation for Design Knowledge Transfer. Unpublished Master's thesis, National Yunlin University of Science and Technology, Yunlin, Taiwan (in Chinese).

Marinkovic, B. (2021). Tacit knowledge in painting: From studio to classroom. The International Journal of Art & Design Education x 40(2), 389–403. https://doi.org/10.1111/jade.12354

Moore, E.C. 1959, "Personal Knowledge: Towards a Post-Critical Philosophy. Michael Polanyi", Philosophy of science, vol. 26, no. 3, pp. 270–272.

Nonaka, I. (1991). "The knowledge-creating company", Harvard Business Review, November-December, pp. 96–104.

Nonaka, I., & Takeuchi, H. (1995). The Knowledge-Creating Company. New York: Oxford University Press.

Özcan, O. & Akarun, L. 2002, "Teaching Interactive Media Design", International journal of technology and design education, vol. 12, no. 2, pp. 161-171.

Özcan, O., & Akarun, L. (2002). Teaching interactive media design. International Journal of Technology and Design Education, 12(2), 161171. https://doi.org/10.1023/A:1015209012458

Polanyi, M. (1958) Personal Knowledge: Towards a Post-Critical Philosophy. Chicago: University of Chicago Press.

Polanyi, M. (1966) The Tacit Dimension. Chicago: University of Chicago Press.

Polanyi, M. (1969) Knowing and Being: Essays by Michael Polanyi. Chicago: University of Chicago Press.

Polanyi, M. (1983). The tacit dimension. Gloucester MA: Peter Smith.

Schindler, J. (2015). Expertise and tacit knowledge in artistic and design processes: Results of an ethnographic study. Journal of Research Practice, 11(2).

Shute, V. J., Sun, C., & Asbell-Clarke, J. (2017). Demystifying computational thinking. Educational Research Review, 22, 142–158. https://doi.org/10.1016/j.edurev.2017.09.003

Smith, E. A. (2001). The role of tacit and explicit knowledge in the workplace. Journal of Knowledge Management, 5(4), 311–321. doi:10.1108/13673270110411733

Smith, K. M. (2015). Conditions influencing the development of design expertise: As identified in interior design student accounts. Design Studies, 36(1), 77-98. doi:10.1016/j.destud.2014.09.001

Smith, K. M. (2015). Conditions influencing the development of design expertise: As identified in interior design student accounts. Design Studies, 36, 77-98. https://doi.org/10.1016/j.destud.2014.09.001

Wong, J., Chen, P., & Chen, C. (2016). The metamorphosis of industrial designers from novices to experts. International Journal of Art & Design Education, 35(1), 140–153. doi:10.1111/jade.12044

Yu, ZH.H.(2002). The debate between scientism and humanism in the perspective of the theory of tacit knowledge. Fudan Journal (social sciences), No. 4. 2002

System Innovation in a Post-Pandemic World – Kin-Tak Lam et al. (Eds)
© 2022 Copyright the Author(s), ISBN: 978-1-032-24392-4

The utilization of pyrolysis products from the waste palm kernel shell

Chien-Yuan Chen*
Program of Mechanical and Energy Engineering, Kun Shan University, Tainan, Taiwan

Huann-Ming Chou
Department of Green Energy Technology Research Center, Kun Shan University, Tainan, Taiwan

ABSTRACT: Waste palm kernel shell was pyrolyzed at 500–550°C in a heating furnace and then cooled to produce the high-value bio-char, biogas, and wood vinegar products. The carbon content of each metric ton of bio-char produced is equivalent to 0.68 metric tons of carbon dioxide, which could be used as a carbon-fixed way to reduce the global greenhouse effect. Biogas had a high calorific value up to 7000 kcal/kg to be as a gas fuel. Wood vinegar could prevent the damage of crops caused by bacteria or insects, and was also improved as a liquid fertilizer by the addition of potassium hydrogen phosphate (KH_2PO_4) in this study. All the products obtained from the pyrolysis were then proved to be advantageous to the recirculation economy.

Keywords: Palm kernel shell, Pyrolysis, Bio-char, Bio-oil, Wood vinegar.

1   INTRODUCTION

Since 2015, the production and utilization of the palm oil has been on the rise. In the process of palm growth and oil extraction, bio-wastes, including palm compound leaves, palm tree trunks, empty palm fruit clusters, palm husks, palm middle peel fiber, and fruit meal, were generated. It was estimated that every ton of palm oil extraction resulted in nearly 5 tons of bio-waste. It was predicted that about 74 million metric tons of palm oil production in the world in 2020 (FAO, 2015) would lead to the generation of approximately 370 million metric tons of bio-waste, and about 10–15% of this bio-waste consisted of the waste palm kernel shell. The accumulation of these wastes may result in anaerobic biological decomposition, leading to the release of a large amount of greenhouse gases such as CO_2, CH_4, and N_2O into the atmospheric air. The production of greenhouse gases could be prevented through utilizing bio-wastes for different purposes. For examples, the oil palm waste could be used as a bio-fuel (K.H. Huang, 2017; H.Y. Zhuang et al., 2018), as an agricultural organic compost fertilizer (K.H. Huang, 2017), and also as derivative biofuel pellets or rods (L. Blair, 2017; Yek et al., 2017).

This paper reported the other ways of utilizing the waste palm kernel shell by generating high-value products, including bio-char, bio-gas, and wood vinegar liquid at a plant (Y.D. Tech. Com. Ltd.) in Taiwan through pyrolysis, and discussed its advantages for the recirculation economy.

2   EQUIPMENT AND METHODS

The plant for the pyrolysis of waste palm kernel shell is located in Tainan City, Taiwan. The plant is spread over an area of about $1600 \, m^2$, which is further divided into a feed storage area, a process area, and a product storage area. The waste palm kernel shells were imported from Indonesia (irregular grains, with length and breadth varying in the range 5–10 mm, as shown in Figure 1), with the moisture content of 11.4%, oil content of 0.44%, flash point of 135°C, dry-state high calorific value of 4608 kcal/kg, sufficient for use as a biofuel, and ash content after complete combustion as low as 2.51%.

The pyrolysis and cooling process, shown in Figure 2, involved a heating anoxic pyrolysis furnace with a movable cover, a water scrubber, and a

Figure 1.   Dried waste palm kernel shell.

---

*Corresponding Author

DOI 10.1201/9781003278474-22

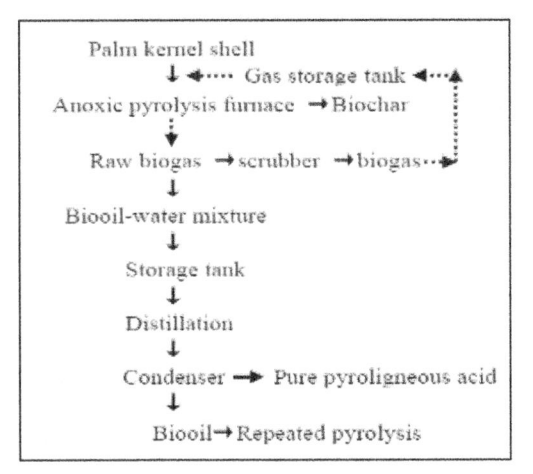

```
Palm kernel shell
        ↓ ◄···· Gas storage tank ◄···▲
Anoxic pyrolysis furnace → Biochar    ⋮
        ↓                             ⋮
Raw biogas → scrubber → biogas··►
        ↓
Biooil-water mixture
        ↓
Storage tank
        ↓
Distillation
        ↓
Condenser → Pure pyroligneous acid
        ↓
Biooil → Repeated pyrolysis
```

Figure 2. Pyrolysis and cooling process for the waste palm kernel shell.
Source: Y.D. Tech. Com. Ltd.

distillation tank. The waste palm kernel shells were first taken by a conveyor belt to the pyrolysis furnace with a volume of about 1 m³ for gasification. The heating zone at the bottom of the furnace was heated by biogas incineration and the temperature was maintained at 500–550°C, which resulted in the waste palm kernel shell gradually gasifying to produce the primary biogas, and, then, carbonizing into bio-char, which was then pushed through the bottom outlet pipe of the furnace by the rotator and dropped into the spiral conveying pipe below the furnace to be finally pushed out of the furnace, and then cooled in the air. The primary biogas produced in the furnace was pumped through the gas pipe on the side of the furnace into two washing towers of volume 55 and 100 liters for condensing, in order to produce secondary biogas and wood tar–water mixture. The secondary biogas was clean bio-gas without ash or particles, which was ultimately pumped into a water-sealed gas storage tank. The wood tar–water mixture was stored in the storage tank and then pumped to the distillation stage to obtain pure wood vinegar.

3 RESULTS AND DISCUSSION

This process used a ton of waste palm kernel shells to produce about 0.28–0.30 metric tons of bio-char (moisture 14.26%), and 0.18–0.22 metric tons of wood tar–water mixture to produce 0.15–0.20 metric tons of pure vinegar 0.02–0.03 metric tons of wood tar after distillation, and 0.48–0.54 metric tons of secondary biogas (water content of about 8–10%). It was reported that the product ratio was affected by the pyrolysis temperature in the furnace (Bruce A. McCarl et al., 2009; E. Jakab et al., 2000) In the following sections, we discuss the application of the obtained products in recirculation economy

3.1 Application of bio-char

The shape of the bio-char particles produced by the pyrolysis process was irregular, with the length and breadth in the range of about 2–10 mm, averaging about 5–6 mm. The quality of the bio-char was investigated by physical and chemical analysis Table 1 shows the average value and two individual analysis values of this analysis. The average moisture content of the two bio-char samples was 14.26% because of absorbing moisture from the air after cooling. However, the average dry-state combustible content was 71.98%, the ash content was 13.77%, and the fixed carbon was as high as 79.99 %, which means that about 80% of the bio-char was composed of carbon atoms or compounds, and the remaining 20.0% of other volatile substances Its dry-state high calorific value was as high as 7400 kcal/kg which means that the bio-char contained a huge amount of combustible components. Bio-char can be used as a soil conditioner to enhance soil's carbon storage capacity, to increase soil water-retention capacity, to enhance soil fertility retention and cation exchange capacity, and to enhance the crop nutrient availability, along with reducing soil nutrient loss, increasing soil microbial activity, improving (increasing) soil pH and enhancing soil aeration (Association of Taiwan Agricultural Science and Technology Resource Management, 2017).

Table 1. Characteristics of bio-char obtained from the pyrolysis of waste palm kernel shells.

Component	Unit	Test 1	Test 2	Average
Moistures	%	14.98	13.53	14.26
Ash	%	10.15	17.39	13.77
Combustibles	%	74.87	69.08	71.98
Volatile matters	%	18.87	21.15	20.01
Fixed carbon	%	81.13	78.85	79.99
N	%	0.40	0.44	0.42
C	%	75.15	73.39	74.27
H	%	1.67	1.62	1.65
Heat values (dry-state high value)	kcal/kg	7400	7400	7400
Space density	g/cm³	0.76	0.72	0.74
Iodine value	mg/g	250	475	–

Source: Y.D. Tech. Com. Ltd.

The average apparent density of the natural charcoal was 0.72 g/cm³, and the iodine values of the two samples were 250 mg/g (unground) and 475 mg/g (100 mesh after grinding), which could be used in wastewater or raw water adsorption applications, or in feed additives and supercapacitor applications. Elemental analysis showed that carbon (C) contents accounted for 74.27%, while nitrogen (N) and hydrogen (H) were only 0.42% and 1.65%, respectively, which indicates that most of the bio-char was composed of carbon atoms. The pyrolysis of 1 metric ton of waste palm kernel shells could produce 0.25 metric tons of bio-char, consisting of about 0.18 metric tons of carbon

(74.2%), which produced 0.68 metric tons of CO_2 when it reacted with oxygen. This bio-char was a product of high-temperature pyrolysis and it would not be easy to decompose the bio-char again at room temperature. If it was sealed underground and isolated from oxygen, it would not be easily oxidized and decomposed by microorganisms. It could be a very stable and effective method for fixing carbon dioxide in the long run

3.2 *Application of biogas, wood vinegar, and wood tar*

After condensing to room temperature, the secondary bio-gas was analyzed by a water flow calorimeter to contain a calorific value of $5300-5700 \, kcal/m^3$, which makes it fit for use as a gas fuel for power generation for plant lighting and running electrical appliances or for other purposes. The wood tar–water mixture was distilled and then separated to produce a small amount of wood tar and a large amount of clean wood vinegar. The calorific value of wood tar was found to be 8830 kcal/kg by calorimeter analysis The wood tar was then moved to the furnace for pyrolysis or fuel. Clean wood vinegar was a natural organic wood liquid with an organic content of about 2.5–6.5% (containing more acetic acid and less phenols) and a pH of about 3.1–4.2. It could be used as an antifungal and antibacterial in agriculture. It is also a natural pesticide (Yongyuth Theapparat, 2018). In recent years, it has been used to produce anti-inflammatory and antioxidant effects in the human body (Z. Rabiu et al., 2020). This study added potassium dihydrogen phosphate to the pure wood vinegar to make it fit for use as a liquid fertilizer and to meet the fertilizer content standards of a liquid fertilizer laid out by Taiwan Government (Council of Agriculture, Executive Yuan, R.O.C., 2000)

4 CONCLUSION AND SUGGESTIONS

Anoxic pyrolysis at a temperature of 500–550°C followed by a cooling process was found effective in cracking waste palm kernel shells in the plant to produce 20 tons of products a day, about 5.6–6.0 metric tons of bio-char 3.6–4.4 metric tons of wood tar–water mixture and 9.6–10.8 metric tons of bio-gas. All the products obtained through pyrolysis had a higher economic value price at more than 40–80 times the price of waste palm kernel shells in the market, which could give a big boost to the recirculation economy.

ACKNOWLEDGMENTS

We are grateful to Yuan Da Technology Co., Ltd. for providing relevant information and investigation data.

REFERENCES

Association of Taiwan Agricultural Science and Technology Resource Management, 2017. Report of International Trend of Biochar Industry.
Council of Agriculture, Executive Yuan, R.O.C., 2000. Liquid Organic Fertilizer (Item No. 5-15), Fertilizer Type, Item and Specification.
L. Blair, 2017. Wood Pellet Market Update, Nordic Baltic Bioenergy Conference, Helsinki.
Bruce A. McCarl, C. Peacocke, R. Chrisman, C.C. Kung, R.D. Sands, 2009. Economics of Biochar Production, Utilization and Green House Gas Offsets, In: Lehmann, J., Joseph, S., editors, Biochar for Environmental Management Science and Technology, London: Earth scan, pp.341-357.
FAO (Food and Agriculture Organization of the United Nation), 2015. FAO Statistics Data 1961-2014, Food and Agriculture Organization, USA.
K.H. Huang, 2017. Resource Utilization Profile of the Oil Palm Biomass, Magazine of Industrial Material, Vol. 372, pp. 038-046.
E. Jakab, G. Várhegyi, O. Faix, 2000. Thermal Decomposition of Polypropylene in the Presence of Wood-derived Materials, J. Anal. Appl. Pyrolysis, Vol.56, pp.273-285.
Z. Rabiu, M.A.A.M. Hamzah, Z.A. Zakaria, 2020. Characterization and antiinflammatory properties of fractionated pyroligneous acid from palm kernel shell, Medicine, Chemistry, Environmental Science and Pollution Research, Published. DOI:10.1007/s11356-020-09209-x, Corpus ID: 218657374.
Yek, Ogawa, 2017. Cost Efficiency of Oil Palm Biomass for Biomass Power, 8th Biomass Pellets Trade & Power Conference, Tokyo, Japan.
Yongyuth Theapparat, 2018. Ausa Chandumpai and Damrongsak Faroongsarng, Physicochemistry and Utilization of Wood Vinegarfrom Carbonization of Tropical Biomass Waste, Chapter 8, Tropical Forests - New Edition, IntechOpen com.
H.Y. Zhuang, Z.X. Xie, 2018. Visiting the Palm Farm and Oil Mill of Sime Darby Company in Malaysia, and the Forest Farms and Plywood Plants of Sabah Forest Industries Co., Ltd., Report of visit abroad, CPC Corporation, Taiwan, R.O.C.

Author index

Smart Science, Design and Technology

The main goal of this series is to publish research papers in the application of "Smart Science, Design & Technology". The ultimate aim is to discover new scientific knowledge relevant to IT-based intelligent mechatronic systems, engineering and design innovations. We would like to invite investigators who are interested in mechatronics and information technology to contribute their original research articles to these books.

Mechatronic and information technology, in their broadest sense, are both academic and practical engineering fields that involve mechanical, electrical and computer engineering through the use of scientific principles and information technology. Technological innovation includes IT-based intelligent mechanical systems, mechanics and systems design, which implant intelligence to machine systems, giving rise to the new areas of machine learning and artificial intelligence.

ISSN : 2640-5504
eISSN : 2640-5512

1. Engineering Innovation and Design: Proceedings of the 7th International Conference on Innovation, Communication and Engineering (ICICE 2018), November 9–14, 2018, Hangzhou, China

   Edited by Artde Donald Kin-Tak Lam, Stephen D. Prior, Siu-Tsen Shen, Sheng-Joue Young & Liang-Wen Ji

   ISBN: 978-0-367-02959-3 (Hbk + multimedia device)
   ISBN: 978-0-429-01977-7 (eBook)
   DOI: https://doi.org/10.1201/9780429019777

2. Smart Science, Design & Technology: Proceedings of the 5th International Conference on Applied System Innovation (ICASI 2019), April 12–18, 2019, Fukuoka, Japan

   Edited by Artde Donald Kin-Tak Lam, Stephen D. Prior, Siu-Tsen Shen, Sheng-Joue Young & Liang-Wen Ji

   ISBN: 978-0-367-17867-3 (Hbk)
   ISBN: 978-0-429-05812-7 (eBook)
   DOI: https://doi.org/10.1201/9780429058127

3. Innovation in Design, Communication and Engineering: Proceedings of the 8th Asian Conference on Innovation, Communication and Engineering (ACICE 2019), October 25–30, 2019, Zhengzhou, P.R. China

   Edited by Artde Donald Kin-Tak Lam, Stephen D. Prior, Siu-Tsen Shen, Sheng-Joue Young & Liang-Wen Ji

   ISBN: 978-0-367-17777-5 (Hbk)
   ISBN: 978-0-429-05766-3 (eBook)
   DOI: https://doi.org/10.1201/9780429057663

4. Smart Design, Science and Technology: Proceedings of the IEEE 6th International Conference on Applied System Innovation (ICASI 2020), November 5–8, 2020, Taitung, Taiwan

Edited by Artde Donald Kin-Tak Lam, Stephen D. Prior, Siu-Tsen Shen, Sheng-Joue Young, Liang-Wen Ji

ISBN: 978-1-032-01993-2 (Hbk)
ISBN: 978-1-003-18851-3 (eBook)
DOI: https://doi.org/10.1201/9781003188513